INDIAN DIARY & ALBUM

CECIL BEATON

Indian Diary & Album

With an Introduction by Jane Carmichael

OXFORD
OXFORD UNIVERSITY PRESS
NEW YORK DELHI
1991

Oxford University Press, Walton Street, Oxford OX2 6DP

Oxford New York Toronto
Delhi Bombay Calcutta Madras Karachi
Petaling Jaya Singapore Hong Kong Tokyo
Nairobi Dar es Salaam Cape Town
Melbourne Auckland

and associated companies in
Berlin Ibadan

Oxford is a trade mark of Oxford University Press

Originally published by B.T. Batsford Ltd. London as Far East, *1945 and* Chinese Album, *Winter 1945–46*
This edition arranged and issued, with permission and with the addition of an Introduction, by Oxford University Press and John Nicholson Ltd. 1991

The collection of some 7000 photographs held by the Imperial War Museum, London, comprises the bulk of Cecil Beaton's work as an official photographer during the Second World War. Visitors are welcome to consult it in the Department of Photographs and prints may be ordered on application.

ISBN 0 19 212299 1

British Library Cataloguing in Publication Data
and
Library of Congress Cataloging-in-Publication Data
available

Designed by Pages Design Services
Printed in Hong Kong by Skiva Printing and Binding Co., Ltd.
Published by Oxford University Press, Warwick House, Hong Kong

Contents

Introduction

WHEN the books *Indian Album* and *Chinese Album* were first published, they were introduced as showing:

> . . . countries, now awakening to a new importance in the affairs of the world . . . seen through the eyes of a great artist with the camera.

For seven months in 1944, Cecil Beaton had been commissioned by the British Ministry of Information to travel as its official photographer in India and China. His photographs, despatched by uncertain routes from the jungles of Burma and the hillsides of China, were used in the contemporary press as part of the official policy of familiarising the British with the more remote theatres of war. On his return, Beaton produced an illustrated account based on his diaries and photographs, entitled *Far East*. But relatively few of the 4500 images he had taken could be included and so in the winter of 1945–46, the two 'albums' were compiled and published. They were at once recognised as containing some of the best of his work.

The Second World War challenged Cecil Beaton as nothing had ever done before. Born in 1904, the son of a well-to-do timber merchant, he had set out deliberately in his early twenties to establish himself as a darling of society, a leading portrait photographer, an original theatrical designer and an amusing writer. It says much for his powers of determination that, by his mid-thirties, he had largely succeeded. However, he was far more at home in the drawing-room than the military barracks, and it was one of the most surprising turns of his career that he should become a noted war photographer. Official photographs were taken under the auspices of the propaganda organisation, the Ministry of Information, and the services themselves, as part of the effort to publicise Britain's cause at home and abroad. A friend of Beaton's, Kenneth Clark, the Director of the National Gallery in London, was involved in the Ministry and suggested that his photographic skill could be useful. Unlike other official photographers, Beaton's work was not submerged in the usual cloak of anonymity; his standing ensured that he received a personal credit and his commissions were specially arranged. In 1940 and 1941 he was busy on the home front showing the engagement of the civilian and the role of the Royal Air Force. In 1942 he undertook his first foreign tour, travelling to North Africa to photograph the war in the Western Desert and to the countries of the Near East to portray their leading personalities.

His photographs and book, *Near East*, met with general approval and prompted the Ministry to plan his most ambitious commission, the journey to India and China to gather material on these little known theatres of war. Between January and July 1944 he covered enormous distances, visiting the Burma front, travelling extensively in northern and central India, and then flying over the 'Hump', the air route over the Himalayas, to Chungking, the wartime capital of Chiang Kai-shek's nationalist China, and from there touring the southern provinces of Szechwan, Hunan, Kwangsi, Fukien, and Chekiang. He swung between the extremes of luxury in the Viceroy's residence in Delhi and the acute discomfort of travel by dilapidated army truck through China. Assessing the experience in his diary he wrote:

> It has been just what I needed, and the toughness of the trip has been beneficial too. For it does no one harm to get tired and to walk too much, to be either too hot or too cold, to go hungry for a few hours. I am heartened to realize how well my constitution stands up to these tests. But I have become painfully conscious of my limitations and mental weaknesses.

But in writing this Beaton was doing himself an injustice; he overlooked the austerity and discipline which the war placed on his photography, stripping it of much of its previous drawing-room glamour and forcing an altogether less self-indulgent and more candid style. Never very interested in technical wizardry, his preferred camera was a Rolleiflex, a twin lens reflex which took two-inch square negatives and was convenient and robust enough to survive the rigours of travel. It allowed him sufficient flexibility to encompass the casual and the formal. Portraiture dominated his work and, where possible, he liked the sitter to pose to his instructions, but his studies such as those of the coolie boys, the merchants, and the schoolgirls in India, or of the destitute, the children and the workers in China, are simple and direct and far removed from the fantastic style of his pre-war days. He was not and did not pretend to be a war photographer in the conventional sense of setting out to compile a visual narrative of the effort and cost of battle. In this sense he was right to recognise his personal limitations, but he overlooked his strengths; his sense of beauty, both physical and natural, his ability to capture the play of light on form and texture, and his eye for a satisfying and serene composition which gave his photographs an appeal which reaches beyond their wartime context. Beaton was intensely conscious of human dignity and his style emphasised the attractive, not the abhorrent. It was an idiosyncratic record of countries at war, but none the less valid for that.

Surprisingly, Beaton tended to disparage his gifts as a photographer, feeling his skills as a writer whose output numbered more than thirty books, and as a theatrical designer whose work included the brilliant costumes and sets for the musical and film, *My Fair Lady*, were somehow 'worthier' than something which came relatively easily. But photography was the consistent backbone of his life. In the late forties he was seen to have matured and although his society portraits were often studies in perfected glamour, he retained the lessons of his wartime work and had the courage to be simple when it suited. His distinguished contribution to artistic life was recognised with a knighthood in 1972 and he died on 18 January 1980.

Later in life he recognised more clearly the importance of his official photographs and described his reactions after a visit in 1974 to the Imperial War Museum where most of them are preserved:

> Looking at them today, I spotted ideas that are now 'accepted' but which thirty years ago, were before their time. The sheer amount of work I had done confounded me.

As interest in Cecil Beaton and the diversity of his talents grows, this timely re-issue of his wartime 'albums' can be enjoyed for the insight they give into his development as a photographer and as a record of his response to the beauty of the countries he visited at a crucial stage in their development.

Jane Carmichael
Keeper, Photographs
Imperial War Museum

January 1991

Indian Diary

To

DIANA COOPER

With Love

Preface

MANY people during the last few years have made fantastic journeys to the far corners of the Earth at record speed. My feats were not extraordinary. But it is possible, I think, that some readers may like to escape with me for a few hours from their everyday surroundings and share the superficial impressions of a traveller who visited for the first time, and in wartime conditions, some quarters of the Orient. I can promise that they will read nothing of politics. Gandhi is not mentioned here. Nor will they find any attempt to solve the Indian problem. Too much criticism has been aired by others — little of it constructive.

This is not an official book. The views expressed are my own. I was sent to India and China by the Ministry of Information as a photographer. It was agreed before I left that any writing I chose to do was entirely my own affair.

I wish to thank all those who received me so patiently and, often in trying circumstances, showed me so much friendliness.

C.B.
April, 1945.

CHAPTER ONE

Departure

DEPARTURE before daylight. Paddington Station was dark, misty and bitterly cold. A group of army officers, some with red tabs, and a few rather seedy looking civilians, assembled in the reserved compartment which would take us to our transport plane. Later, in an enormous Nissen hut, on a moor in the most inclement corner of England, we awaited news from the meteorological experts, who for days had been watching the movement of a depression. There were patches of snow on the ground. As they huddled around a sulky stove, sailors, who knew the North Sea, said they had never experienced such weather. The pilots were joking among themselves: "I haven't been so cold since I spent New Year's Eve in jug." A line of nearly twenty transport aircraft, the overcrowded Mess, the presence of Field Marshal Smuts, trying to warm his calves at a fire in an anteroom — all indicated that this was a serious hold-up.

For eleven hours we waited. Our pilot, a young blond Canadian, lean and lithe, with a stance like a gorilla, had high cheek-bones and pale almond eyes that closed upwards when his mouth twisted into a smile. He edged towards the stove to rub his hands. At two o'clock in the morning, in another icy hut, we had breakfast — an egg and bacon and coffee. "Smuts has taken off," they told us. Now there was a chance of our leaving in a couple of hours. But my spirits were low. My whole body ached with the cold; and I was full of apprehensions.

Zero hour — "Yes, we're off."

We were led out into the sharp blackness of the night and dimly lighted to a lorry. First stop, a farm building on the moor where we were trussed up in boiler-clothes that seemed to add bulk without giving warmth. The interior of the aircraft, when we reached it, proved to have been stripped of all but the minimum equipment. One side of the fuselage was packed with luggage and miscellaneous cargo, huge rubber tyres for airplanes, crates and "secret" packages. The door was locked. We sat in thick blackness and listened to the roar of the engines. Then the aircraft trundled forward, bumping along the uneven frosty ground.

The agony of terror that followed, though it lasted only a few minutes, seemed an eternity. Already, at the start of the run, I bowed my head in my hands and prayed very hard because I was so frightened. Then my terror was intensified. My eyes were shut tight, and I tried not to take cognisance of anything outside my own head; but somehow I felt that these were my last seconds of life, and I decided that I must spend them contemplating pleasant subjects. All sorts of unexpected and forgotten pictures raced through my mind. I saw my family when I was a child. I saw a schoolboy named Geoghegan, waiting for me to finish my school tea, as was his custom, under an arcade of chestnut trees, outside the playground of my old preparatory school: he waited to give me a lift home on the step of his bicycle, and our friendship was tranquil and harmonious. I relived my sensations of excitement produced at Christmas by the gift of a picture postcard of a musical comedy favourite. I had idyllic memories of the first time I fell in love — of a soft welcoming gesture of affection from my small house in the country during the height of summer. I remembered the gaiety of certain New York winters, and again smelt the hotel-rooms I occupied. I had visions of the silver grey trees, against blue skies, that Piero della Francesca painted in his frescoes at Arezzo. . . . This was all too pleasant: beauty doesn't consist only of pleasure: and I tried to make myself visualise uglier things; but I could think of nothing that was not ecstatic. My ideas worked up to a crescendo of clear vivid thought. I was in a delirium of pleasure and terror. Outside my own thoughts I knew that my deepest fears had turned to reality, that my anxieties had become actualities. In the air our machine was banging violently from side to side. "Yes, that's it. Now we're for it!" I heard someone say.

I opened my eyes: through the cracks of the closed door leading into the cockpit I saw flashes. Suddenly we were rattled like dice, and I found myself lying on a mound of parachute harnesses, halfway down the fuselage. Now the flames were everywhere. A huge tongue of blue flame darted the length of the cabin. The cockpit was an orange glow. Outside the night was lit by enormous different coloured fires. In the aircraft were odd patches of flame, and a bright phosphorescent fire centred in the extreme rear. The engines continued to roar: and I supposed that we were still airborne. So this was the end; this was death; any second now I should know the unknown. Meanwhile I analysed quite calmly the various stages through which I passed. No use fighting death — there was nothing to be done about it. The flames approached. Everyone was very quiet in the aircraft, and even now behaved with the polite reserve of Englishmen. I looked up to see the whole fuselage illuminated by a dense suffocating orange smoke, through which the silhouetted figures of the air-crew, in their cumbersome divers-suits, ran past, groping in the fog of burning aluminium. Still no one spoke. I lay holding my head, thinking that, as soon as the flames reached us, there would be panic and fighting and I should be trampled underfoot. And why not? I had accepted the worst; this was it. Somebody shouted: "Open that bloody door!"

I could see various passengers hopefully and pathetically groping for an exit. One of the civilians had the presence of mind to turn his torch on to the latch of the door. Its beam seemed very white in the glow of the fires. Then I understood, by some queer reflex, that the door was open. "So they are jumping for it," I thought, "rather than be burned. How high are we? Well death is one stage further this way. . . . So here goes!" I crawled along the floor backwards and tipped myself out, head first, into the cold black night. A short drop; and I was astonished to find myself, with a minor bump on the head, upside-down in a grassy field covered with hoar frost and patched with snow. The air struck me as bitterly cold. Around and above me were flames.

"Get up and run," someone shouted. "The airplane may explode." In spite of a tremendous weakness in the knees, and the weight of my cumbrous clothing, I ran, as we all ran, falling, getting up again and running, turning at last to watch the destruction of the plane from the vantage point of safety. The broken monster lay spurting flames. Deep ochre and black smoke coiled upwards in a great tower: the cockpit was diamond bright; the burning edges of the wings suggested flare-paths on an aerodrome, or gala illuminations of a pre-war pier. Our lungs filled with fumes, we coughed as we watched. It surprised us to find how little shocked we were. Someone said the shock would come later. It did. Meanwhile we gazed at the burning plane, as it vomited out different coloured flames and spat forth distress signals of pink, mauve and golden rockets. We discussed our miraculous escape. We had crash-landed — another fifty yards and we would have plunged into the ocean. What was the cause of our crash? An engine iced up? But no one knew.

I could not feel proud of the negative way in which I had behaved: just to lie and accept death was of little help to others. . . . It was a passenger pilot who had known how to pull up an emergency lever and jettison the locked door.

"Are we all here? Are you all right?"

The airfield was dotted with flame-lit figures. "The pilot didn't get away," remarked the navigator. Fumes brought tears to our eyes as we looked at the funeral pyre of the charming young Canadian. The night wind was icy and cut the scalp like a knife. Eventually the ambulance came up. And then, thank God, staggering out of the darkness, his neck and forehead bleeding, his face green, appeared our pilot. He had been thrown clear of the aircraft a hundred yards away. He was taken off in the ambulance suffering, we discovered later, from internal injuries. He had a broken arm and ribs; a kidney had to be removed — the stick had gone through his stomach; his flying days were over.

Accepting the fact that we were safe, each of us now remembered his particular treasures as they burned before our eyes.

"There goes all I possess. I've nothing but what I stand up in," the passenger pilot said. "Most of all I mind losing the photographs of my child. They were taken at different stages ever since he was born."

"All my papers have gone," said another man. "The result of weeks of meetings."

"I've got my bag all right," the deaf Wing Commander, who was travelling as R.A.F. Courier, added. While the others had been fighting to get the door open, he had been throwing the baggage about to find his precious burden. In the lorry, on our way to hospital, someone congratulated him. "What I want to know," he replied, "is, will they give us another breakfast. I'd go through that again any time, so long as they give us another egg."

CHAPTER TWO

Delhi

THE bearer, white turbaned and bare-footed, pulled back the curtains to let in a blaze of sun. Outside, the fountains were playing, the birds shrieking: someone was practising on a bugle, and sentries cleared their throats with a resounding rasp, spat and stamped their bulbous boots on the gravel. Another bearer, in scarlet tunic, came in to salaam and give me a present. Yet another servant, in an enormous cheese-cloth puggari, brought in a necktie wrapped in coloured paper. It was Christmas Day in New Delhi . . . .

My second attempt to fly to the East had been uneventful. On my return to re-equip myself, after the crash, London had seemed grimmer and more forlorn than I had remembered. The weather was consistently grit-grey; there was an epidemic of influenza; everyone seemed un-nerved; my resistance was low and I chafed against the tyranny of small restrictions. My second set of travelling companions behaved like all travelling companions. Wherever we landed we ate gluttonously. In the air we complained of the unaccustomed surfeit of food, and proceeded forthwith to help ourselves to a meal out of a carton and some soupy tea from a celluloid cup. By the end of each day, the places we had left behind were remote and dreamlike in our memory. Before each dawn came the hangman's Reveillé, followed by exhausting hours of doing nothing. Occasionally the voracious silver porpoise would alight to refuel. Rather crumpled, the passengers emerged from its entrails for a breather in an oil-smelling launch. At last, in the unbecoming heat of the midday sun, we saw the great brick blisters, pig-pink and opaque as plasticine, of New Delhi.

My first days were spent wandering down the long corridors of the Secretariat. Arrangements were made for me to go to the Burma Front, to the North-West Frontier, to Madras, to Kochin. But each day brought some alteration of plans, some delay or disappointment; and I soon understood that I would have to stay longer in Delhi than I had intended. "This is Headquarters from which all arrangements are made," I was told. "But, old boy, you can't expect anything to be done overnight. It all takes time. You should have warned us before. You see, *your* trouble is, old man, that you come under so many different headings! You see, there's S.E.A.C.: there's H.Q. India Command (that's us): there's the Government of India, the Far Eastern Bureau, the Ministry of Broadcasting and Information — there's .... But I'll take you to Brigadier Oldfield of S.E.A.C. — he'll help you."

It was difficult to hear quite what was happening in this crowded small office. On the telephone Major Arnold was giving someone hell for spelling Air Marshal with two L's; while from outside came a fearsome noise as of souls in torment — a dozen natives, unsuitably draped, were trying to lift a safe. Unperturbed, Brigadier Oldfield planned an itinerary for me on a map. "Then you go to Cox's Bazaar — or Bawli Bazaar — get a plane at Ramu for Chittagong — on to Camilla. Let me explain," he pointed, "this is the front here — we're moving towards Maungdaw."

I was allowed into the War Room of South East Asia Command. The chiefs of all departments, American and English, "breezed in" for early morning prayers, to study the latest maps, and to hear the day's reports and short lectures given by half a dozen specialists. The Supreme Commander, Admiral Lord Louis Mountbatten, who had arrived in this theatre not long before, seemed as yet unaffected by the climate. "We mustn't let it be a damper on effort — we've got to galvanise everyone, got to teach 'em to hustle," he said — and he appeared to have impregnated his immediate entourage with his own robust enthusiasm. In spite of all the difficulties he had encountered, no glaze of disappointment was visible in his eyes. They twinkled with the delight of a boy who has just been given a Meccano for Christmas — which incidentally, I believe, was just about all he had been given. For was it not decided, at the Teheran Conference, that the Eastern theatre could not be a scene of great activity until the European war was over?

In the War Room, sitting among admirals, air-marshals and generals, the Supreme Commander would interrupt the lecturer to ask pertinent questions. Mountbatten was ebullient: his toy seemed to be working well. But some of his minions had gone to sleep again. It was early morning still; and the droning voice of the lecturer, in the otherwise silent room, acted as a soporific.

Indian flautist

Early morning lecture to the leaders of South East Asia Command

While awaiting further instructions I spent many days sight-seeing as far away from Headquarters as possible. I was excited by the glimpse of my first wild parrot, monkey, elephant, and stimulated by the brilliant, poisonous colours and ceaseless movement of Old Delhi. In the Chandi Chowk (the Street of Moonlight), at one time considered the richest street in the world, now an alleyway full of bargains and trash, a begging Sardhou, naked and daubed with dung, an "exponent of destitution", extended a withered arm. Other holy men had whitened faces; and there were boys with heavily kohl-painted eyes, their teeth, tongue and lips scarlet with betel nut. In the thoroughfares, pedestrians, bicycles, carts and sacred animals were wedged together in an almost inextricable confusion. Women resembled human bee-hives, entirely covered with white cloth except for the small letter-box slot through which their painted eyes peered. The shops, no more than window-recesses, contained spangled tassels, glittering phials of perfume, filigree jewels and vivid foodstuffs. A stall of vile-coloured drinks, in bottles stopped with fans of magenta paper, had been built around a sprawling Peepul tree; its bark, painted emerald green, gave the scene an added gaudiness. Everywhere one found unexpected sequences of colour — no gorgeous oriental riot but an extraordinary jumble of apparently inharmonious colours, harmonised nevertheless by some unknown law. Wonderfully lovely, for example, were the Rajputana women's skirts of dirty tomato red, with dull powdery rose or lobster pink draperies. White played a large part in the general scheme; and the touches of scarlet, such as the broad border on a child's white coatee (worn over white trousers) or the alazarin cape thrown over a shoulder, were tremendously effective.

In the neighbourhood of Delhi, I visited early Hindu forts of the Eleventh and Twelfth centuries, and Early Mohammedan forts and cities. Sometimes the day would end as we wandered among the Pillars of Victory, the shrines, the relics of the old cities of Delhi, or lingered at the fort of Tughlakabad or the Tomb of Humayan. From the parapet of one of these great monuments, in the precious moments of twilight, one saw India at her best. In the foreground, the domes of mosques, originally bright with coloured marbles, but much more beautiful now in their present quiet tonality: in the distance, the lilac-coloured jungle. A crescent moon would appear, in silvery contrast to the few wisps of golden cloud that were hurrying to be away before the sky became completely dark: cranes and other large birds were flying home and their wings made a breathless flapping noise: while parrots, very small, but tightly clustered, gave the impression, as they passed, of a flying carpet. Jackals came out and slunk off again, horrible hangtail scavengers. A shepherd, rather sadly, was playing on his flute; and from the distance we heard the echoing call to Evening Prayer.

One afternoon, all the way from Old Delhi to the Safdar Jung Tomb beyond the new Capital, the highways were filled with a great concourse of Mohammedans, taking part in the yearly festival of the Mohorrun. The crowds on foot, or brimming over the sides of bullock carts, were in their best clothes. In the West, people seem to choose colours for no particular reason. Here each colour appeared to indicate an uncompromising personal preference. One woman was a walking rainbow, in a small crinoline of apricot yellow that faded to pink and mauve. A ragamuffin had staked all on a surprising dark-red coat of the finest quality velvet. A delicate looking little boy wore, very correct and straight across his brow, a gold embroidered cap of deep grape colour; the cut of his tight-fitting coat and the precision of his taste were such that one felt this perfect work of art should have been preserved in a showcase. Down the streets, enormous edifices of coloured paper and tinsel were borne on poles. And each flimsy temple represented a very definite taste: one, of orange and silver, seemed to be conscious of its loveliness; another, of white and pale pastille green, was timid and tentative. Each had its own variety of rhythmic movement, as it swayed or jogged along under the dark trees. On a piece of high ground, parched and pale yellow, with gnarled trees and rocks, the procession halted. The paper edifices were savagely pulled to bits, soused with water, then buried in a muddy grave of wet sand. Circles of wide-eyed spectators watched some formalised sham-fights. But what these strange sights symbolised, I had no idea. It was enough to watch. . . . In a haze of churned-up dust hawkers sold bouquets of magenta and white paper roses and brilliant striped sweets. Orange coloured curries were being cooked over glowing cinders.

Here was much for the eyes and much for the imagination. I was constantly amazed, for example, by the beauty of the people themselves. Women's faces, peeping from tinselled draperies, reminded me of doves; their bodies were as compact and firm as bronze statuettes. The good looks of some of the men seemed almost alarmingly arrogant; but others, oblivious of their haunted, haunting beauty, could not understand why a European should wish to stare at their lank hair like the foliage of water plants, or the extraordinarily aristocratic distinction of their limbs and features. The squatting positions they assumed, knees drawn up to chin, as they rested or meditated, reminded me of the bird world.

More delays. Much brilliant work comes from Delhi. No finer example could be found than that set by the Viceroy. He is a paragon of truthfulness and deliberate fairness, industriousness, executive efficiency and courage. The Indian Civil Servants have an uncanny sense of political reality; there are many anonymous, but none the less important cogs, without whom the wheels could not rotate, who toil goodhumouredly in the face of much opposition, criticism and the disadvantage of a thoroughly noxious climate. New Delhi has come in for a great deal of opprobrium; brickbats aimed in certain directions are apt to hit the wrong target. There are, in Delhi, certain strata of people, individually unimportant, who by their frivolity and thoughtlessness create, especially among the men who are fighting in the awful conditions of the Burma Front, a certain rancour. The cry is heard that, despite the activity at Headquarters, Delhi is too far behind the front line of War. Yet it is unfair to expect those who have not lived in the fighting areas to act with the same determination as those who have. All the aids to escapism are available in Delhi. There is small chance of a flying bomb; European food is plentiful; no shortage of manpower, servants galore, countless boys to preserve the tennis court and pick up the balls for the players, masses of old men to water the herbaceous borders, which are a blaze of dull pink, bright orange, butcher blue — a seedman's triumph perhaps, but utterly hideous. There is little noise and the lack of traffic, except for the tinkles of bicycles at luncheon time, gives an air of leisure and prosperity.

Inside many of the freshly painted villas, one might be transplanted to the Great West Road, with nice easy chairs covered with chintz and a bunch of Cape Gooseberries in an art-pot on the modernistic tiled mantelpiece. There are no beggars or shoe-shiners — India is banished successfully, except for the crocodile of ants which winds itself along the bathroom walls, and the mongoose running across the drawing-room floor. Many an officer or government servant and his wife live here in greater comfort than that to which they were accustomed at home. In Welwyn Garden City they would have rallied with their neighbours to keep a stiff upper lip, and their tails high in the face of odds: it is merely pathetic if, after a long sojourn in this Oriental Garden City, they are almost driven flesh-potty. In such an erosive atmosphere it takes a strong man to maintain his balance and a sense of perspective.

Some glib criticism has been tossed around about the state in which the Viceroy's House is said to be maintained. However I was impressed by the style and economy in which the great household not only conforms to wartime standards but appeals to the Indian love of "Bhari Tamasha" (meaning "grand display"). Arriving from England in the depths of the fifth winter of total war, it was extraordinary to see such sun, glitter and colour, so many flowers. I was immensely struck by the presence of so many servants of different categories, in scarlet, white and gold liveries, standing about like poppies behind chairs and tables, or in the distance of endless halls and marble enfilades, looking as small as "personages" in a landscape. But soon I realised the Comptroller had the "Bhari Tamasha" well under control.

In the Viceroy's House three hundred servants are employed. When considering this number you must realise that, due to the caste-system, different communities are allotted various categories of work, and that it is impossible and irregular for a man to substitute for another; thus six servants are needed to do the work undertaken in England to-day by one heroic aged peeress. Any Englishman, living however quietly and simply in India, will have at least six servants — a cook, a butler, a laundryman, a sweeper, a groom, a gardener and perhaps one other. Even so he will be poorly attended, his bungalow dirty, food badly cooked; each servant, willing to do only one specific job, is inadequately trained and incompetent. If an Englishman is to work hard in this devitalising climate he must preserve his energy and leave his servants to do some of the physical work he would undertake in England.

The pretentious buildings of the Viceroy's House and the Secretariat are of no known style. Made of tongue-coloured stone, which retains the dry heat of the day and throws it out angrily at dusk, they appear, at the far end of a processional drive, like a city built for an exhibition. They were designed for peacetime activities, but few modern cities could be less practical or convenient for a war headquarters than the present Capital.

These domed and turreted buildings have had jerry-built beaver-board excrescences added to accommodate, by day, the vast staffs who, by night, must sleep in tents, which, five minutes after a tropical downpour, are flooded, or, in the heat of summer, resemble a furnace. Living space is so scarce in New Delhi that even the migration of South-East Asia Command to Ceylon did not appreciably alleviate the discomfort. The long, empty avenues and spacious vistas may be advantageous for a processional drive of the Commander-in-Chief and his wife, accompanied by outriders, staff cars and police, but the ordinary soldier with a staff job, journeying backwards and forwards to his desk four times a day, dislikes these distances. Even to-day there is no adequate tram or 'bus service to Old Delhi; those unable to find accommodation in the New Town must cover the distance of seven miles by bicycle, for a tonga moves almost as

Imperial Delhi

Moslem and Sikh troops taking the Oath

slowly as the proverbial bullock-cart, while the cost of a taxi is out of the question.

Hours of work throughout the year are long, for no siesta is encouraged in Delhi. By the end of the hot season, tempers are short and nerves are frayed.

The general effect of New Delhi is of a complacent yet callous centre, without gaiety or the strength for cruelty; a heartless, bloodless Display-City, without a past or the necessary roots to develop a future.

After the usual dawn delays at the airport, I set off for the North-West Frontier. We landed at a small place named Cheklala.

Darkness was now upon us; the weather had "closed down" for further flying. Here we must remain. I was a stranger. No one knew of my arrival — yet, in a few seconds of landing, with that extraordinary generosity shown to outsiders by the R.A.F., I found myself whisked off for a drink in the Mess, and forthwith taken to the home of an unknown young man, who made himself responsible for all my needs. He turned out of his room so that I should spend the night in the more comfortable bed. I was dosed with his precious whiskey. I was motored to Rawalpindi, introduced to his friends at the club, given the best dinner I had eaten since the war began (a steak as thick as a dictionary), shown the local belles in full evening dresses at a Charity Dance — in fact, was presented to all the glitter of this famous Station, now used as a training ground for India Command. Next day, many and various methods of instruction were displayed. Gurkhas, in crash helmets, were doing a course of parachute landing: the aircraft circled over a ploughed field: one by one the umbrellas opened in the sky: they appeared suddenly like frog-spawn or undersea life. The stiff figures swayed from side to side as they descended, like those dreadful dolls that used to hang in the back-windows of the automobiles of travelling salesmen. A somersault — they disentangled themselves from their skeins and staggered through the mud for a mug of tea at the mobile canteen.

At Nowshera I was impressed to see one dozen English officers mostly in their twenties, in the process of training three thousand Sikhs, recruited from the neighbouring villages. It was interesting to see the latest arrivals, very willowy in their dhotis and puggaris, and to compare them with the stalwart batch that had arrived only seven days ago. In one afternoon I watched the classes of men at all stages of training. Some were boxing, the onion-shaped knob of hair that their religion demands coming down during bouts and causing the contestants to look like Edwardian ladies; boys were throwing grenades, working on transport repairs, using bayonets, and finally, after a year's training, were at the Passing Out Ceremony, incredibly efficient, swearing the

Oath of Allegiance on the Koran Sharif, or the Sikh Bible, carried aloft on purple and yellow cushions.

Everywhere I visited I was welcomed in such a friendly manner that I should feel ungenerous were I to give the impression that I was disappointed in my first fleeting glance at this remarkable corner of the Earth. I met men whose life had been spent here, and others who resented being stationed so far from the front line. A General, who showed me much hospitality, and who had been through the fiercest fighting in Burma, lost all his possessions there, and was fortunate to have escaped, said he regretted being considered too old to continue the fight against the Japs. "I would like to fight them," he told me, "knowing for once that the airplane above my head may be ours and not theirs." He chafed against his bars in this remote prison, and was disappointed to find here so much apathy, local gossip and pettiness. One young officer confided that he spent his spare cash as conscience money on sending books to the troops in Burma.

The town of Peshawar has been sacked so many times that nothing of architectural interest remains. But the streets are crowded with interesting types — Asiatics, Pathans and many Persians. The shops are a series of enthralling, enlarged peep-shows. The fruit-seller kneels on his prettily built structure of brilliant materials — oranges, melons, magenta aubergines. The hatter squats among the bead and tinsel headdresses, made especially for a bridegroom. The florist is busy stringing garlands of white, peppermint-pink and orange flowers for a woman's wrist or ankles, or for a horse's head. Bright and spectacular shops display dazzling jewellery, brilliantly coloured bed-posts like toys, Ali Baba pots of brass for incense or pungent and warm perfumes, and stolen goods. The grain shops look like miniature models of mountain ranges. Most mysterious of all is the "flour-sifter" on his white stage; he wears a white smock and a white dunce's cap, his face, beard and eyelashes are powdered white, his sieves and strings are covered with a frosty film, and he stares back amusedly as we gaze at him as if he were from some other world.

My escort, perhaps in an attempt to create an air of excitement, explained that, if the police should turn its back for ten minutes, the place would be in an uproar; that here we were surrounded by a fermenting mass of the world's most dangerous characters. That evening, however, everyone seemed agreeably inquisitive. "Of course it seems quiet here to-night," said my guide. "But you never know when it will be necessary to turn on the Tear Gas again. This is a tough corner of the Earth. No value is given to a man's life. You notice everyone carries a gun; robbery, hold-ups, murder and rape are not uncommon." He spoke of the "possible danger" with a certain relish.

"You see that type — there?" he added, pointing out a rather pathetic look-

Gurkhas

Frontier guards

ing man who was deliberating in front of a sigillated tea-cloth. "Well he's one of the worst types — bloodthirsty scoundrel."

In high expectation, and in a torrential downpour, I set off for Landi Kotl and the Khyber Pass. Scenery was on a gigantic scale, without a blade of grass; beneath lowering skies these slatey mountains looked even more formidable. The Fort Shagai, housing the Second Kashmir Infantry, combined for me all the least attractive features of a soldier's life: early calls for parades on the asphalt yard, draughty bare rooms, hard gritty ugliness. Alexander the Great and Timur the Tartar had chosen this path for their invasions of India; I felt nevertheless that the Khyber belonged rather to Kipling than to any earlier period of history. "Victorian improvements" had given it a Boer War aspect.

A battalion of the Seventh Rajput Regiment went off down the hills, towards Afghanistan, on a tactical exercise in frontier warfare. Everywhere one moved one was watched by pickets looking down from camouflaged pillboxes on the mountain heights. I was depressed rather than impressed, and perhaps I did not show enough enthusiasm; for my escort continued: "You must understand what a poor life these tribesmen lead. They see, next to them, the most fertile plains of all India, yielding four crops a year; they cannot help coveting such richness, and they make continuous short, sharp sorties to grab a bit of someone else's wealth. The hostile tribal territory here is always a problem; and the terrain makes it impossible to winkle them out of their caves without an enormous expeditionary force."

"Perhaps they are rather a good buffer between India and Afghanistan?" I suggested.

"Oh, they provide us with a lot of fun. It's a great life — hard, but it's masculine: not a woman in sight."

My friend then told me about the Fakir Ipi, who, as a young clerk, was sacked from a Government Office and ever since, inspired by a fanatical hatred of Britain, has been making trouble for us. For many years we have tried to capture him. We know where he is, but he never comes out into the open; and none of his followers has been found willing to deliver him up. Although he has been quiet lately, trouble is always expected. "He's got his own arms-factory underground, you know; and no doubt he is paid by the Axis for kicking up shindies and keeping our troops busy on the frontier. But he's over fifty and suffers from asthma, so perhaps it won't be long now. By the way you heard about Richardson?"

"No, what happened to him?"

"Well, he was sitting by his window — Phft! The light was shot out." My friend explained how a wild tribesman had missed his mark. Pickets were sent out; extra care taken by the police; scouts sent to protect the traffic on the mountain road from guerillas who might be lurking there with homemade rifles. There might be some subsequent potting from behind boulders.

Later, I saw a group of the wild Wazirs, whose activities had at one time considerable nuisance value. I have seen "Carmen" and "The Maid of the Mountains" on tour, and can recognise a third-rate chorus of operetta brigands. Here they were again, after all these years; with unkempt beards and dirty undergarments swathed round their shaggy heads. One wore a long green tweed overcoat of loud check with emerald celluloid buttons, obviously bought in a bargain basement: another sported an old tail-coat. Toothless, squinting, stunted, with inane grins — a more hopeless bunch of delinquents it would have been impossible to imagine.

I went down the "Open Road" guarded by the Tochi Scouts. I watched guards signalling from slope to slope and saw, on the peaks of these gaunt hills, white sheets placed as indications to aircraft. I saw the Scouts, with the agility of goats, scaling in thirty-nine minutes a height that would take a white man two and a half hours to achieve.

Back in the Officers' Mess, polished, silver cups stood in rows against the dark oak panelling. Another round was ordered — "Yes, we get beer from the factory at Pindi — or how about a cherry brandy?" A young subaltern came in and laid his revolver on the table, by the reading lamp with the crimson silk shade. "Heard about old Claude's near shave? His lamp shot to blazes! Great stuff — maybe the beginning of something."

Life on the North-West Frontier has changed very little since the Victorian age, when warfare was so well-conducted as to seem comparatively civilized. The subnormal mountaineers are still a nuisance; but, after all, they are the inspiration of a thousand mess-room stories. A hundred years ago, this Frontier possessed a romantic quality, which it has largely lost since the invention of more modern forms of frightfulness — the flame-throwing tank and the flying bomb.

It was time to leave. We called upon the Station Commander. No, he did not think there would be a chance of the aircraft arriving in such bad weather; certainly none of its taking off. But while the meteorological people were making gloomy forecasts, explaining that this greyness might keep up for several days, the duty pilot announced the arrival of our aircraft. We motored out on the tarmac. "Okay, we go in half an hour," the little Indian pilot shouted.

Again I was the only passenger in the aircraft, and concentrated very hard on my novel as we rose into the pewter skies. We bumped about in a steely vacuum, passed through the storms and came out into ambiguous calm. I soon noticed, however, that the pilot was trying very hard to unwind a wheel — something to do with pumping down the under-carriage when the automatic release goes wrong. It proved too stiff; try as he might, he could not get it down. Now we were flying over nasty, toothlike rocks, and into large lumps of dirty cotton-wool cloud. The little Indian was sweating as he struggled with the levers. Since my crash I have lost my former sublime confidence in the infallibility of aircraft, and am conscious of the slightest irregularities in the rhythmic sound of the engines. When the pilot first beckoned me to join him in the cockpit I shook my head and winked — No, I had had enough of the cockpit: I would remain with my novel! The pilot continued to beckon; and it was only after a considerable time that I understood that the invitation had now become an order. Imagine my dismay when I discovered the pilot was signalling for me to sit by him in the cockpit, to "take the stick". I had never been shown what to do with the controls. I felt like Harold Lloyd flying an airplane for the first time, and contemplated even having to climb out on the wings, to tie something together with string. I held on to the wheel rather gingerly, not knowing how much leeway I could allow before the aircraft reacted violently. Like a monkey, the sweating pilot crawled to and fro, among the hundred gadgets on the dashboard and the floor. The engine responded to my very tentative suggestion to climb a little higher, and I found this effort a relief.

After an eternity, the pilot put up his thumb with a jerk: he had mended the airplane: perhaps he had tied it together with string and a safety-pin. "Okay."

"May I go back to my novel?"

Thumb up again. The relief was tremendous. When we circled over Lahore Air Station, however, it seemed the thumb-jerk had been premature. As we came in to land, and were just about to touch down, we shot up again high into the air. The under-carriage was not lowered. We "stooged" around the air-field, while the pilot tried to unwind the under-carriage. He kept re-adjusting fuses: we circled many times, looking down wistfully at the strip below. The pilot took control again. Perhaps he had decided to "do a pancake" without the landing gear? Here goes. I looked around for the quickest means of escape — in which direction would I be thrown? — and adopted several suitable poses in which to receive the shock. But all went well. We skimmed low, bumped, and were relieved to find the under-carriage was in position. Only the wingflaps were not working, so that our speed was greater than usual, and again and again we bounced high like a rubber ball. But we were safe on land. I had been so frightened that I felt as if I were suffering from prickly heat; my hands were sweating, and I had a painful weakness in the small of the back.

CHAPTER THREE

Calcutta

CALCUTTA, the second largest city of the Empire and the former capital of India, is an ugly, dirty and unhappy city. There are many splendid parks, tropical gardens, squares adorned with Edwardian statues, and florid commercial edifices of unknown styles; and there are many enchanting eighteenth-century buildings. But the overcrowding, the poverty and shabbiness are distressing. Only fifteen paces across the main thoroughfare, opposite the most elegant European hotels, village-like groups are clustered around the fires, over which heavily spiced food and bits of fish are being fried in grease under the trees, while hordes of rats scurry to and fro, and scavenger dogs and enormous crows take an interest in the refuse bins. Ladies in opalescent evening dress, each at the beginning of the evening the proud possessor of a well-starched escort, take themselves to the Philharmonic Concerts on Sunday evenings, where the Indians present a dazzling display of jewels. A few hundred yards away, at the Kalighat, the most primitive scenes of worship in all India are to be seen. Calcutta seems doomed to disaster. The climate, perhaps the most unhealthy of any town in India, may be responsible for the weakness, indolence and apathy of so many of its inhabitants.

It seems that during wartime the most agreeable (and the easier) way of attacking the British Government is *via* Bengal. To launch a more direct onslaught might be considered ill-timed and unpatriotic. But Bengal is a permissible target, and a large number of strangers have acquired a militant determination that England shall alone be made responsible for disasters which are not so much the result of short sightedness and callousness, as of a series of separate mishaps, uniting to aggravate a situation already complex and difficult. The visitor to Calcutta soon becomes aware that the recent famine gave opportunities for a shocking display of avarice, selfishness and almost inhuman apathy among the people themselves. It is not easy to help someone who will not help himself. One need only read the descriptions of Indian famine written in the eighteenth century by William Hickey to see that the Bengali mentality has not changed, and that such wholesale tragedies may periodically recur.

"Nothing," wrote Hickey, "would stop the unhappy famished wretches from rushing in crowds to Calcutta, the neighbourhood of which became dreadful to behold. One could not stir out of doors without encountering the most shocking objects, the poor starved people lying dead and dying in every street and road. It was computed that for many weeks no less than fifty died daily, yet this patient and mild race never committed the least act of violence, no houses or go-downs were broken into to procure rice, no exclamations or noisy cries made for assistance; all with that gentle resignation so peculiar to the natives of India, submitting to their fate and laying themselves down to die. Everything in the power of liberal individuals was done for their relief: indeed, one must have been less than mean, absolute Bonapartes, to have witnessed such horrible scenes of misery without feeling the bitterest pangs and exciting every nerve to alleviate them."

To-day Calcutta, so recently recovered from famine, is thriving. Fortunes are being made. Directors of firms are "reserved" from the army, and in the end, no doubt, receive knighthoods. Calcutta is known as the city of "dreadful Knights". Moreover, it is now a sort of oriental Clapham Junction. Air Commodores, Generals and celebrities of every kind and race spend the night at one of the overcrowded hotels or at the vast Caravanserai of Government House. Most of the men of the Fourteenth Army spend their leave here. In their bush-hats and shorts they crowd out the hostels, canteens, air-conditioned cinemas, cafés, milk-bars and "attractions" along Chowringhee. It is a thrilling experience, after years in jungle foxholes, to walk along stone pavements, to gaze up at tall jostling buildings and to sleep all night in solidly constructed edifices. After listening to the whispers of the jungle, the violent noise of the town comes as a relief. Calcutta provides plenty of noise: the tick-ticking and thunder-rolling of the trams, the honking of taxis — for Sikhs always drive with the horn — the bells of the rickshaws and the incessant caw-cawing of the crows which circle above.

The crowded streets provide every sort of contrast. Poles wander in search of distraction; American sailors, with cigars at an insolent angle in their mouths, buy silk kimonos embroidered with dragons, or the Taj Mahal in sequins; British

Tommies rather clumsily finger the bookstalls for semi-pornographic literature: there is a choice between "The Seven Pillars of Foolishness", "Gone with the Monsoon", "The Art of Love" or "Erotic Edna".

In the markets are to be found all the oriental junk that Birmingham can produce; carved ivory by the ton, engraved metal and elaborate enamelling. Rare animals of the jungle are brought together under the glass roof of the market: caged birds of all colours and sizes, and pathetic monkeys in crates. Some black, long-haired monkeys have eyes like wallflowers and the dignity of saints. A young boy, holding a birdcage, pauses a moment to rearrange his coloured girdle; another is sitting upon a trestle, sharing it with a goat and many vegetables; a young Hercules saunters by, balancing an enormous wardrobe on his turbanned head. One forgets how beautiful the human body can be until one sees it with the draperies so enticingly arranged.

To walk down the tiered flower-stalls is as exciting as a play: roses of varying charm and personality; bundles of white daisies like clouds; the first mauve orchids. There are flowers virginal or mock-virginal, some suggesting muslin, others confectioners' sugar. In the secondhand market you find eighteenth-century furniture brought out by the sailing ships, and Victorian bedheads, covered with a thick coating of treacly gilt, or with the glutinous mahogany polish that the Indian loves so much. The sign painted above the fronts in Bow Bazaar are often startling: "Specialist in Piles", "Specialist in Wet Dreams". There is much evidence of the wish to become white-skinned — creams are advertised that will make you like a lily.

Thousands of worshippers, night and day, are busy in the humid turmoil, purchasing silver-paper treasures, praying or washing themselves in the holy waters near the site of the Temple in Nonsur, of the Goddess Kali, the wife of Siva, or festooning white flowers over the ugly images. A goat is blessed; holy water is poured over it, and other attentions are paid: it is whisked to the stocks; its head clamped it gives forth its last terrified bleats. Drums roll. With a terrific swirl the knife descends. The head is severed from the body; both lie twitching and spurting blood. The crows come down quickly to peck but they have little luck, the blood dries very quickly in the sun, and soon becomes a black stain on the paving.

Around the Kali Temples, the sacred waters, the tree of fecundity, Miss Mayo has done her job. Perhaps she has not stressed the great beauty of the scene, blazing with aniline colour during the day, phosphorescent in the moonlight. In the midst of its squalor, Calcutta provides strange and unexpected glimpses of beauty. In the ugly heat of the day, a young man, in rags, sits in an alcove of the wall outside Government House, oblivious of the raucous noise and the clanging traffic, idyllically playing a lute. In the "Victoria School", the young girls look like little statues, their heads bent forward and palms clasped as they make obeisance to the teacher. I have never seen so many exquisite, kitten-like creatures gathered together in one room, each with her own brand of beauty, but all distinguished by the same magnolia texture of skin, velvet eyes and lacquered simplicity of hair.

Adjacent to the tramlines, the cemetery, the tombs and mausoleums of the eighteenth-century English pioneers, is a most hauntingly beautiful and strange confusion of Georgian and Oriental architecture, obelisks, pyramids, temples, among them the pathetic resting place of Rose Aylmer, with huge birds flying above, and frogs croaking underfoot.

The great Jain Temple, in a compound containing a series of architectural mad-houses, is the apotheosis of bad taste. Brought from far and wide to be jumbled together, is a preposterous collection of objects, *bric-à-brac* European and Oriental. The result is defiant and amusing. In the brilliant sun, statues of potentates and ugly children, in staring white marble, with black inlaid pupils to their staring eyes, stand up under pagodas of painted wire, on mosaic-work pedestals. There are wrought iron garden chairs from Llandudno, Majolica elephants from Birmingham, many turquoise-roofed temples, all of a Turkish Delight prettiness, pavilions that might have come from Brighton Pier inlaid with coloured glass, stones and mirror work, like the most expensive nougat. The whole effect is as gay as the tinkles in the Czechoslovakian chandeliers: a suitable setting for an Opera Bouffe.

On the banks of the Hooghly River there is a piece of land endowed for the Sadhus. Having given up all worldly possessions, these holy men satisfy their frugal wants by begging, and cover themselves with an ash that contains a sulphur which makes their naked bodies impervious to changes of heat or cold. They practise Yogi, and each morning go down to the Hooghly to bathe and do their muscular exercises, using the river water to irrigate their bowels. The fact that they spend most of their time smoking hemp and hashish does not affect their physique; the bodies of even the older men are wonderfully lithe and energetic. Some of them laugh mischievously, peep and leer around corners. The young men with their long bleached hair hanging below their shoulders, naked but for a scarlet jockstrap, with their skin powdered half-elephant, half circus perfomer, look more like devils than holy men. One naked young man, with his wild hair flowing behind him, comes charging down to the water astride a great bull. An old man with one eye, one tooth, and hair growing in plaits yards long, reads from the Holy Book. He is the head Sadhu. To bask and bathe in his blessed presence many visitors from the town have come to sit cross-legged and silent in front of him. A nine-year-old-boy, with grey powdered face and hair, in scarlet draperies, looking like an angel painted by Signorelli, sits in the Lotus position, playing an enormous musical zither twice his size. This child has a remarkable passivity that is very moving. His dank, dark and dirty surroundings make him appear even more strangely pure and beatific. For the atmosphere is faintly vicious and sinister, though maybe it is only the feckless, rather uncanny, laughter of the youths that give one that impression. The old man explained that it was unfortunate we had elected to visit him on a day when so many of his number were away: twenty of them had gone to Nepal. I did not realise that this meant that they had gone begging their way on foot, and that it would take them a whole year to reach their destination.

I began to realise the size of the country known as India. Place the map of the Peninsula over that of Europe and one discovers that from Delhi to Calcutta is the same distance as from London to Danzig. From Delhi to Bombay is the same as from London to Rome. From the snowy heights of the Himalayas southwards to the ocean is a distance of two thousand miles, and the width of India from the Khyber pass to the Eastern border is nineteen hundred miles.

Bengal is the largest jute producing country in the world. The raw material is stacked in thousands of aerodrome-hangar-like storehouses. A dozen Indians grunt and groan, or shout as they load the heavy cubes on to the backs of two human runners. The cubes are then dumped on to a small railway train which is driven into the factory. A jute mill is a gratifying sight that leaves nothing to the imagination, for one sees all the various stages by which the finished product is achieved; and the different scenes are pleasing to the eye. Even the jute itself is a sympathetic substance, like pale pussy-willow-coloured silk.

I had heard that the Scots overseers were hard task-masters and I had imagined that the conditions would be appalling; but, although the noise was tremendous, a dry peppery smell everywhere, and the air filled with fluff, the atmosphere seemed cheerful and sympathetic. Many of the workers are from the country. While they are working in the factory they lead a bachelor life in barracks, but for three months of the year they return to their wives. To follow one piece of jute through the factory is a most illuminating experience. After the long horse-tails of fibre have been whisked into a rubber gyrating trough, they are waved into tresses, like marcelled hair; next they are whirled around on spools like tops, then threaded into a spider's web of strings. Now they are jerked into a violent loom, from which they emerge as immaculate lion-coloured cloth. Men, with elaborate nether draperies, fold the cloth into giant dinner-napkins. These are stacked, cut and sewn into sacks by squatting women.

Rows of women, heavily jewelled and in draperies that run the gamut of wine colour and amber, line up like caryatides, their stitching finished and the sacks on their heads, to deliver their work. The sacks are clamped in metal strands; they are packed tight, and sent up in cranes over the Hooghly river to be dumped into the boats below and taken to the most distant parts of the world, to be used for packing food or making sandbags. More recently, these factories have provided the tents that protect the army on the Burma Front.

In a neighbourhood of cheap modernistic apartment-houses, of honey-comb tenement-buildings that seem so unsuitable for the climate of India, I visited the house of an Indian poet. I was struck by its peaceful beauty and simplicity; the atmosphere was one of an extraordinary sweetness and purity. Transparent dhotis and white saris, freshly laundered, were hanging from the landings and balcony, as emblems of cleanliness: the golden ewers sparkled in the washroom: the stark, almost empty bedroom, with the poet's children asleep, was innocent of all the unnecessary and stuffy impediments of a humble room in the Western World. Here were no cluttered drawers, here the essentials alone; and yet, when I wanted, of all things, a tripod, the poet was able to produce it.

Off a side street, in his studio, sits Jaminy Roy. He looks like a long baked potato, nestling in a napkin, in his immaculate white muslin. In front of him, on the floor, are many bowls of different colours. He paints as if he were making decorations on pottery. Once he was an academic portraitist; but Jaminy Roy was dissatisfied with oily, fulsome likenesses of rich people that he was able to reproduce with facility and technical skill. He retired to a small village, studied Matisse and other modern painters, made his own water colours and started to paint in a brilliant and vital manner. After many years of poverty and hardship he is now considered India's best modern painter. His pictures are bold and simple. But he produces too much at too little cost to himself.

In a gun and shell factory, where supplies are made for the war in Burma, the atmosphere is little different from that in any such factory throughout England: it is as clean, no more, no less, the manager as impersonal. But the Indian workers seen better suited to precise and continuous tasks than workers of a more volatile and restless temperament. One sixteen-year-old boy was at work planing and filing a small part of a gun; the delicacy and sympathy of his hands, the coolness of his gestures and the meek intensity with which he laboured, were very touching to watch. The employees here

Wedding-cake architecture: Jain temple, Calcutta

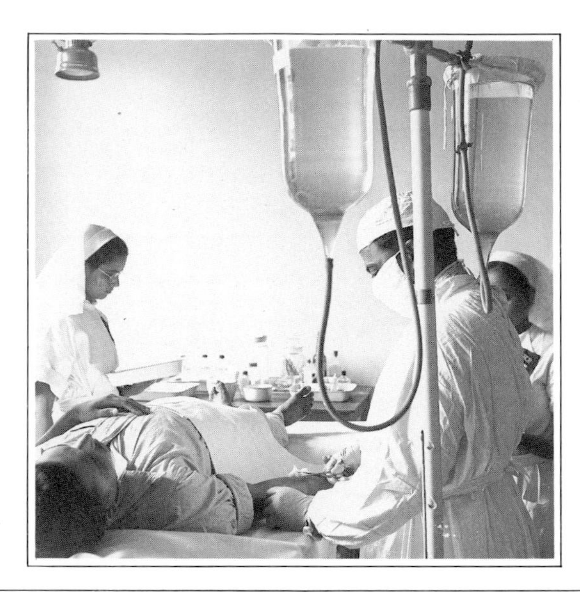

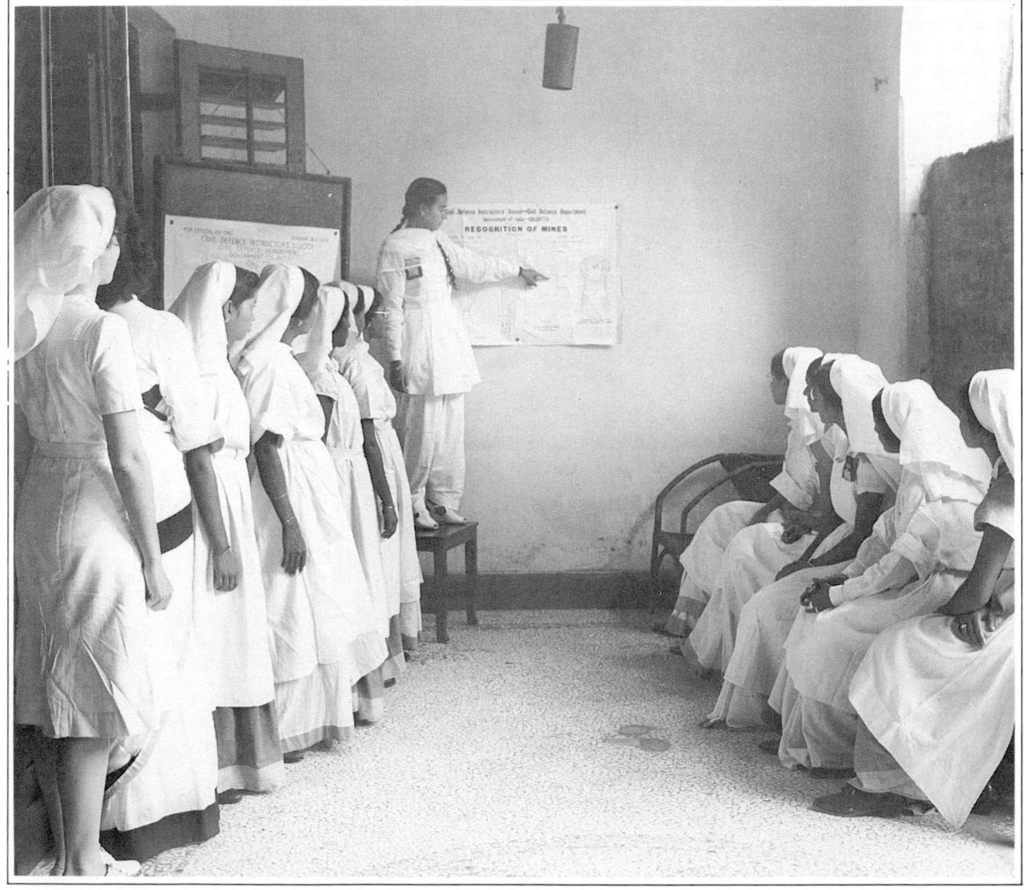

Calcutta hospital and A.R.P. lecture

are given periodic propaganda talks, but few of them have any hatred for the Japs, or any real idea about the war. Most of them are working because they are paid good money. As workers, they do what they are told with prodigious powers of application. It is doubtful whether their efficiency would be greater if they had the cause at heart. In another huge factory, in the heart of Calcutta, precision instruments, optical lenses, telescopes and gun-sights are made. Some of those employed here are fourth-generation employees, descendants of those who have been associated with the firm since its inception at the time of the East India Company.

Shell factory, Calcutta

The Governor had prorogued Parliament. There had been such disorderly scenes in the Legislative Assembly that it was decided to curtail them. I went to hear the final flurry.

An "Alice in Wonderland" mad-house presented itself. Everyone at the same moment seemed to be shouting and beating the air. One man, with a voice like a siren, moaned above the others, demanding an opportunity to speak without interruptions. I noticed later that when others were speaking, he was the first to take the chance of raising a good old shindy. The speaker, like a Grandville drawing of an insect, had a hard time trying to keep some semblance of order. He cried into a microphone, banging to no avail with his mallet. A dignified man, in a tarbush, kept shouting, "Mr. Speaker, may I go on? Oh, they won't listen!" he wailed. "They don't want to hear truth and correct information — please Mr. Speaker, oh please prevent us from becoming a laughing stock."

It was impossible for a stranger to follow proceedings. The metallic voices jerked out their English sentences in a timbre that was difficult to listen to; but the use made of the English language was refreshing.

A fat old man in a dhoti rose to his sandals and shouted: "This is most vexatious for the Honourable Minister!" Others took up the cry, "Vexatious, most vexatious!" Finally the minister, who was supposedly so vexed, rose and remarked deprecatingly, "I can assure you it is not vexatious. I am not easily vexed."

The Speaker gave hopeless rulings. The document from Government House, proroguing the Assembly, was greeted with shouts of "Ignorrrit — ignorrrit". The bedlam of noise and confusion rose to a crescendo, to be ended abruptly by the Speaker adjourning the house for fifteen minutes of prayer.

No ship bound for India sails straight to Madras or Calcutta; it calls first for orders at the embarkation port of Bombay. The gateway of India is the first glimpse that our troops have of this great peninsula. What can be their impressions? Is this India? The turreted square might be part of Liverpool or Edinburgh; these once prosperous looking red brick buildings are enlargements of Pont Street houses; the Town Hall, a classic edifice, might be seen anywhere throughout the world; and the rows of modern apartments, along the reclaimed land of the Back Bay, look like accumulators and have no nationality. The Taj Hotel is built back to front, and in the moonlight, with the electric lights reflected in the calm sea, it is reminiscent of Buda Pest. There is the Sassoon Library. There, reminiscent of the Great Exhibition, is the Crawford Market. Those imitation Trianons are the mansions of rich Parsis. There are old-fashioned clubs and modern hotels, synagogues, mosques, Buddhist chapels and the Towers of Silence where the vultures silently receive their meals at punctual intervals. Along the walls of the town, perched like monkeys on scaffolding at various heights, are Bombay's never idle artists, perpetrating huge garish posters that advertise anything from Nautch dancers to snow-white skin pigmentation. These adroitly painted faces of gargantuan proportion and cineraria complexions are one of the town's most personal features. Below, dwarfed by comparison, the crowds include women in brilliant saris, pince-nez, brown boots, and American and English sailors in white. Religious signs are painted on Hindu beads in the gutter. In the shade of a parasol attached to his white tunic, the policeman beckons to the oncoming traffic in a dashing uniform of navy-blue plus fours and yellow porkpie hat.

The Exhibition Stadium changes its attractions with necromantic speed. No

The Royal Indian Navy

Bombay rice supply

sooner have the stalls been taken away from the "Grow More Food" exhibition, than the crowds arrive to watch the wrestling matches. Then, in a flash, a transformation takes place and the attractions are installed for a huge Red Cross Fun Fair. Allied service men, W.R.N.S. in tropical uniform, taciturn Indians in turbans and white Eton jackets, and Maharanees armed with diamond bracelets to the elbow, rubicund members of the Yacht Club, wearing white dinner-jackets with black trousers (whereas their counterparts in Calcutta wear black jackets and white trousers), enjoy the glow of coloured electric light bulbs, the gambling, the donkey races, the roller-coaster railways and the giant wheels which revolve high among the fronds of palm trees.

In another part of the town, forbidden to the troops, in rows behind their iron bars, the ladies of the night display their charms and to the highest bidder extend the key of their illuminated cage.

With its vast dockland, its miles of godowns that contain equipment of war, Bombay is the store-room for India Command. The Military Ordnance Depot is an incredible sight, with acre upon acre of tanks, armoured cars, DUKS and — perhaps equally precious — rubber tyres. Bombay is not self-sufficient; but its food problem has been tackled by an alert and far-seeing Government, who were the first to institute rationing schemes. In spite of the illiteracy of so large a proportion of the populace, control of grain, sugar and kerosene runs smoothly, without queues or disorder. Attached to the large factories and cotton mills are rice stores which are models of their kind. The mill areas stretch for miles into the country. Forty years ago almost every cotton garment worn in India was woven in Manchester; now Bombay supplies the goods. I am certain that the Indian mill-hands work in better conditions than their equivalents in England. Women workers can leave their children to be looked after in a cool, immaculate crêche in the care of ayahs, until they visit them at feeding time.

Less artificial than Delhi, less dirty than Calcutta, beautifully situated on the sea, Bombay cannot be considered an Asiatic City; but it is a throbbing Eastern metropolis that welcomes Western civilisation. It is the most cosmopolitan and emancipated city in India. In spite of its orchid house climate, its inhabitants seem to possess unflagging initiative and drive. Sects, clubs, associations and newspapers are legion. Bombay is also a great town for gambling — particularly among the Parsis. Many of those present at the race meeting each Saturday have dreamt about Doubles. "Of course the favourite will win," someone in the crowd is heard to say, "or the stewards will want to know why. . . ."

As the horses flash past, the crowd groans in a vast orgasm of excitement. The heat is terrific: a few of the young women are extremely decorative in their clear coloured saris; some of their menfolk, with tweed jackets worn over their muslin shirts, look messily indecent; but the general effect — the bright, coarse flowers set in stiff borders, the whites, greys, bright salmon and mauve pinks, with a distant rainbow in the sky — has the period charm of a Tissot painting.

On the far side of the white railings, a great altercation results from somebody not having cheated as much as he expected to do: the row is brought to a conclusion by a violent downpour of dramatic rain. In a flash the crowds have dispersed — not before getting soaked through. The drainage system does not allow for such a rainfall. Lawns are flooded, cars are waterlogged: a few straggling Indians paddle with battered umbrellas held aloft in one hand, shoes in the other; and husky B.O.R.S., like children at play, proceed by slow degrees, climbing along with their stomachs pressed to the railings.

CHAPTER FOUR

Assam, Burma and the Arakan Front

WE landed in the bowl scooped between the mountains of Imphal. The year was at its best; sun all day; cold at night; the cherry trees in in blossom, rhododendrons ablaze. Soon there would be orchids; the vast tropical trees would be transformed by these exotic parasites, and the troops would pick the blossoms and put them in their large brimmed hats. The impressions I received would have been far less pleasant had I arrived during the Monsoon period, which continues for nearly two-thirds of the year.

Living in small holes dug in the mountain sides, supplied by a narrow mule track which zigzags up and down the mountains for over three hundred miles from the nearest supply base, the men must exist soaked to the skin for weeks on end in an almost solid tropical rain. There is no chance of drying their clothes. In this fetid atmosphere to wear a mackintosh is to sweat so much that soon you are wet through. Boots are never dry, so that your toes begin to rot. Supplies suffer: the coarse flour breeds bugs. Mud reaches up to the thighs. Everything grows mouldy: even the bamboo poles grow internal fungus, and the smell of decay is everywhere. Transport becomes impossible and essential supplies have to be dropped by air. Yet, strange as it may seem, water is often short — the mountains are so steep that the rain shoots off the sides before it can be cupped — and washing is permitted only once in three days. The enormous trees, garlanded with festoons of moss, drip heavily, ceaselessly, for months on end. Mosquitos thrive, and the leeches appear in millions, wagging their heads from side to side in the elephant grass. They are small until they have feasted on human blood, when they swell to the size of your thumb. The soldiers have learnt that they will drop off if touched with a lighted cigarette; but, if you try to pull at their greasy black skin, the head remains embedded in your body and the wound becomes septic.

Jungle warfare, consisting as it does of lonely treks and skirmishes — at the most, men go out in twos and threes — demands the highest degree of courage on the part of each individual. Most men prefer desert warfare, although here there is shade, the roots and growths are a salutary substitute for fresh vegetables and a palatable addition to iron rations, and occasionally there is wild game. But the feeling of loneliness is greater; groups seldom trespass on one another's terrain. There is reassurance to be gained from fighting in numbers. Each man knows that, after a terrifying game of Blind-man's Bluff played through the coarse undergrowth, any encounter may end with a clash of knives. No quarter is asked or given. Every moment of the day each man must be on the alert; for the Jap sniper may be hidden behind that distant cliff, or in the nearest tree. There is the continual strain of listening for the sound of a footfall. Even during their sleep most men keep one ear open for sounds of the night. They develop a sixth sense, so that they can distinguish every animal step, the calls of the birds, the laughter of hyenas, the yells of jackals, the creak of bamboo, the snapping of a twig and the Aristophanic chorus of frogs and crickets. After a time, even the most robust may show signs of nervous stress. One man, hearing steps coming closer to his Basha, ran out in the dark and bayoneted a bear.

A man is lucky if he escapes for long the various fevers and sicknesses of the jungle. If he should become ill while on patrol behind the enemy lines, he is a liability to his fellows, and a strain on morale. Fortunately the possibilities and probabilities are not considered — "You don't think about anything but killing Japs. It's important to kill them, because then they can't kill you." On fronts nearer home the wounded have a chance of being repatriated; in Burma the wounded can hope, at best, for a bed in a Calcutta hospital. Yet many of these men have been "holding the fort", almost barehanded, for three or four, or even seven, ghastly years on end. Some of them were in the terrible evacuation from Burma. Most of them are hardened, trained, jungle fighters; and they know that, with their experience, they are not likely to be replaced, that they must bide their time for a decisive campaign, which is not likely to be fought now till the European war is won. Yet they grumble surprisingly little. Perhaps they are past grumbling. Many of them had become cynical — "We're just the blokes that are taken for granted." They have received little attention from the outside world; their activities do not make news. They know that, in comparison with the bitter fighting in other theatres, their job is on a small scale, can be considered one of endurance.

"What do people think at home about this particular war?" they would ask with wry smiles, "or don't they think?" — "Do they know we are fighting the Japs out here?" — "We're very far away. But have they forgotten us entirely?"

They had a reason for feeling neglected. Up to that time they had received little recognition from press or radio, and few visitors; the mail from home was spasmodic; and although the Viceroy himself has always made it his business to see that there should be no mail hold-ups, letters still do not receive priority, and the bags would remain for days en route while other cargo was unloaded. The mail from the United States, with an extra two thousand miles to travel, arrived with almost incredible expedition.

The arrival of the Supreme Commander, driving his own jeep, dispensing with all formality and showing a genuine sympathy for the troops, produced a rise in morale. Though incapable of "talking-down" to anyone, his spontaneous speeches were couched in the terms the men appreciate most: "I want you to see my mug — not that I think it's a good mug — but you must get to know your leader. Some of you chaps have been out here for three years — I've been out here for three weeks — and I, too, want to get back home."

Here are some pages from the diary that I kept:—

"We were awakened in the dark; shaving in a small basin in a cold semi-outdoor room was depressing. We started off for Tiddim in a fifteen-hundred-weight lorry. The hearty onslaught of the captain of our party, so early in the morning, was hard to bear. At breakfast he whistled through his teeth in imitation of a tram conductor, did other impersonations, and on the road shouted abuse in four-letter English words and in Urdu to our fellow-travellers. I have never seen a display of such physical energy sustained for so long.

"Our truck bounded about in a cloud of dust thrown up by the convoy of trucks ahead of us. The beauty of the tropical vegetation through which we passed was ruined by the coating of salmon-pink dust, churned by ceaseless traffic. The bamboos, their fronds of dead branches looking like fishing-rods, rose in a perfect pure arc. The high Peepul trees were festooned with ropes and garlands of other vegetation. After a couple of hours the real bumping started. These trucks are the least suitable vehicles for negotiating narrow ridges cut into the precipices of the mountain sides; but there was no jeep available. For hours we were tossed from one side of the truck to the other, thrown high in the air to land painfully on the little iron seat, or on the sharp edges of our baggage. When we went over especially big bumps, the luggage and tinned provisions were thrown with us. We bounced and bounded for the rest of the day, wondering if it were possible to survive so long without being sick, suffering an appalling headache, loosening some teeth or acquiring internal troubles. But the flesh is very strong. With only an interval of half an hour for a lunch meal, we continued in semicircles up or down a mountain side, over a surface of dust and potholes, for one hundred and sixty difficult miles.

"By degrees the sun had warmed the icy cold air; one side of the mountain became brilliant, the other half remaining black. Then the sun sank behind the hills where the Japs were in occupation, and everything became pitch black. Still we motored along the small ridges, past perpendicular drops of four hundred feet; sometimes a passing lorry scraped our mudguards. We stopped to be given a cup of tea, brewed on his primus stove, by an old man in the Signal Corps from Merseyside. He welcomed our company, and his revolting brew was hot and invigorating. He said Christmas out here had been lousy. They had been given breakfast in bed! But the cooks must have become over-excited, for lunch had not been ready until a quarter to four: then it had consisted of just the usual rations — some bully beef wrapped in pastry, but a double ration of rum. He had got his letters from home fairly promptly; but he was lonely here, and had been out too long.

"Our captain showed his exhaustion by shouting at passers-by even more violently. He had taken on the Herculean job of steering this heavy lorry around hundreds of hairpin bends throughout the day, and like Hitler's, his patience was now practically at an end. We barged, crashed, thudded, ricocheted on into the night. The mountains were dotted with the small glowing fires of native encampments. After many dark vicissitudes, with distant lorries approaching like glow-worms, and passing us in a crescendo of noise and blinding light, we at last arrived, under a starlit sky, on the top of a precipice covered with fir trees. We did some unpacking, sat over a fairly warm fire, and waited while the sure, but very slow, black servant prepared tea and sardines, and unrolled our beds."

"The captain awoke very early. From the moment he opened his eyes he was in tearing spirits and carried on an incessant ribald conversation, while I lay counting the number of silly schoolboy words in each sentence. When a patch of sun was reflected on to the wall by me, I felt things might improve. The sun soon filled the hut and took the chill off.

"I went out — with cramp in the neck as a result of a night spent lying on a wooden chair — into the frosty, brilliant blue mountains. Below, the procession of lorries was bringing up provisions in an unending stream. More provisions were being dropped by airplanes flying only a few yards above the fir trees. Boxes swayed down on diminutive parachutes, and bags of rice and

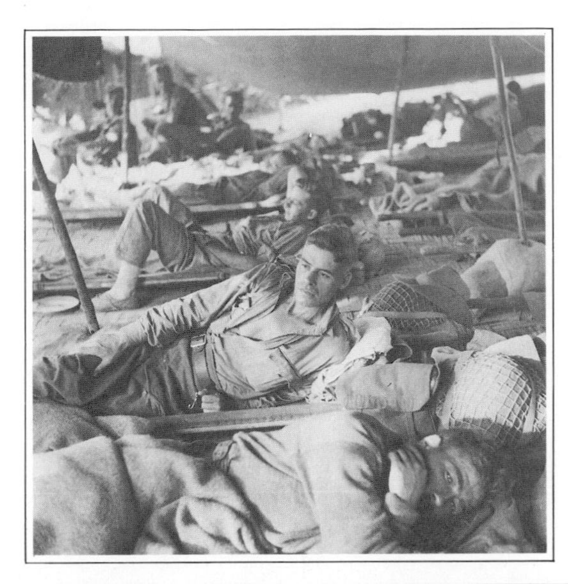

Front-line hospital

Tropical kit

Jungle cookhouse

Howitzer firing

arta (coarse flour) came hurtling through the air. Some of the Chins ran in alarm. Quite a lot of damage had been caused by these supplies crashing through a roof; two days ago a Chin was squashed flat.

"We started off to Divisional Headquarters, to the encampment where five thousand people live in rush-matted tents. Already their toilet had been disposed of; they were slick and polished as if for the parade-ground; shoes shiny, everyone immaculately shaved.

"Against the green surroundings of the jungle the face of a white man is seen easily at a distance; so faces are 'made-up' with dappled spots of blue and green grease. Corporal Mitchell, from Perthshire, looked like the original Bairnsfather 'Ole Bill'; in spite of his maquillage, he had carefully waxed the ends of his large moustaches. Tin hats are worn with sprays of tropical leaves threaded through their netting cover. The white turbans of the Punjabis are veiled with layers of coarse camouflage net: Sikhs appear as if from a ballet, their turbans covered with huge woolly tufts of green and blue; and the Gurkhas patrolling with mobile wireless sets, tall branches like wings on their shoulders as they lean forward to penetrate the undergrowth, look like tropical Burnham Wood on its way to Dunsinane.

"The British gifts of improvisation had been fully exploited. Typewriters were buzzing, and the most elaborate systems of telephone and wireless were installed. But living conditions were almost savage. At night the men slept in fox-holes dug into the peat-like earth. The khaki dhobi (laundry) festooned the branches of the trees; the 'furniture' was made of the strangest objects, and the whole picture was reminiscent of a boys' adventure book. Everyone, young clerks and grey-haired Brigadiers alike, wore short trousers and romantic looking bush-hats. Everywhere was an ant-like activity. Up before sunrise, after working at highest pressure all day, men often find another batch of work that must be completed after the evening meal. This helps morale. At a place so remote — it is a ten days' journey to the nearest town — there is little else to do. Everyone was extraordinarily cheerful, though it was almost more than they could bear to ask for news of England. 'What's the blackout like?' — 'Do they have enough to eat?' — 'How's the bomb damage?' — they enquired rather shyly. When I told them that only five weeks ago I had been in England, they eyed me as if I were a creature from another planet. They touched my civilian jacket and said: 'Can't remember how long it is since we've seen tweeds.'

"After nightfall I sat in a cavern dug in the earth in front of a blazing fire, talking to Colonel Younger, one of the men who helped to build the mountain road over which we had travelled.

" 'It's a promenade now, compared to what it was a few weeks ago,' he said. 'While building the road, we lost only one jeep over the cud. Since the bulldozers have arrived, one of them went over the edge and the driver lost his arm, but the work has galloped ahead. The precision of their work is extraordinary. Those bulldozers do each day as much work as fifty Chins, though it is difficult to aggregate Chin manpower with women and children included. I'll take you up to Kennedy Peak tomorrow; I'd like to show you the flowering trees on the way.' He talked about the flowers and the orchids as if he were showing me around his garden in Sussex.

" 'We mustn't be late for the guns,' he said. We ran in the dark up the mountain side. It was cold at the outset, but we arrived at the summit out of breath, panting in the unaccustomed altitude.

" 'Two minutes to go — one minute to — half minute — FIRE!' A twenty-five pounder gun let loose eight rounds. The noise hurt. It brought to the surface all the soft places in one's body — the places where one's teeth had been filled — the nerve centres and the dormant fibrocitis in the nape of the neck. The noise hurt the ear drums. The blackness of the night became vivid with the flashes.

"Even while fighting a most primitive form of warfare, far from civilisation, men on the Burma Front contrive to lead the life of cultivated persons. I had an agreeable dinner in 'A' Mess. A small flickering lamp does not encourage reading; but a group of men will sit up late in serious discussions that may continue for many nights on end.

"Early call: the misery of shaving from a tooth mug. We motored higher up the mountains in circles to see the 2.5 Gurkha Rifles, a Battery of 129 Field Regiment R.A. These guns occupy the highest ack ack gun site in the world according to the men manning them.

" 'Must be pretty nippy at night, isn't it?'

" 'Thirty degrees of frost,' said Gunner Donovan of Hammersmith.

" 'What do you do about it?'

" 'Put up with it.'

"From Kennedy Peak a panorama of a gentian blue mountainous world lies below. A little way down the slope a young man sat with his Brigadier. He was reporting his experiences of the last seven days of patrol on foot behind the enemy lines. He pointed to the Japanese emplacements and, like all others here, knew the country so well that he was able to describe the distant mountain formations as if they were his homeland. 'See that spear there — that hump — that bare patch — that saddle — that strip of straggly trees — that long cloud formation?' And the Brigadier followed the directions through his field-glasses.

"We started off on the trip down the mountain. The Provost Marshal had misinformed us about the timing of the convoy's departure. We found ourselves in a gigantic crocodile of trucks that were to accompany us throughout the journey home. One truck would break down: a halt along the roadside for all the others: a start: another halt: then another halt in the stream of traffic. It was impossible to pass on the narrow crags overhanging precipices. We waited in the ever-increasing dusty, dry heat. By degrees the freshness and resistance were sapped from the body. Soon one was tired, and not cheered by the discovery that, after six hours, one had travelled only thirty miles. The captain's temper was an incentive to me to keep calm; with his efforts at the wheel of the truck I could only sympathise. The mere physical exertion of steering the wheel round the hairpin bends, the sudden stops and starts on the knife-edge precipices, with a drop of one thousand feet over the cud, was a terrible strain on nerves and physique. We trickled along the passes at a rate of five miles an hour, if we were lucky. In the increasing heat we would stop again for twenty-five minutes. On for five minutes, and then another twenty minute stop. One became obsessed by the problem of whether a far distant truck was or was not on the move. Although we did not give up trying, we could never pass any vehicle. In the distance, on the mountain above, the line of toy trucks would move along at a slow but steady pace, aggravating the chagrin of those stuck below. When, at last, we moved on, it was now at a rate of only three miles an hour. We were maddened to find that some soft-hearted booby in front was pushing on his bonnet a broken down truck, with the result that hundreds of vehicles behind must follow his pace. We managed to send a Military Policeman to bawl hell out of the Good Samaritan. We had still one hundred and thirty miles to go on appalling roads with incessant bends which our truck was unable to manoeuvre without at least one reverse.

"How did one survive the eternity of the next ten or twelve hours? I learnt about the captain's life. To understand him was to make allowances. I had hated him bitterly at the outset of our trip, but my heart warmed towards him when one afternoon I found him poring over a map with his moth-eaten old servant, giving a lesson in geography. The old Chin had no idea which shape signified India, or where Burma was. He looked at various maps and made hopeless guesses. 'This is India. Where is Bombay?' The dark finger would point to the centre of the Peninsula.

"The captain had been a Regular, wounded in an arm and leg by the Japs; had been towed across a river in a net, which was kept afloat by empty bottles; had walked out of Burma on foot with a dozen others; nothing to eat — most of them had died from exposure and starvation on the way. Now he wished to return to his regiment, but was unfitted, and consequently soured; for soldiering was the only profession he knew.

"Mountain scenery never appeals to my heart; but I forced myself to enjoy the various textures of the earth, the stubble of the trees in the middle distance, the silky blue, lake-like distance. I watched the pleasing calligraphy of the various tree branches. In order to pass the time I tried to train myself to see the passing trees as certain painters would have conveyed them — Grünwaldt — Altdorfer — Corot — Samuel Palmer — Dufy.

"We went on and on, without a drink and without food. My thoughts would be far away when, suddenly, I would be brought back to earth by some obscene observation made by my companion. Later, in our grimy exasperation, the scenery appeared detestably dust-covered. At twilight we stopped by an Indian Detachment, dipped our mugs into a cauldron of tea and were given chips, fried in sizzling grease. I don't know if the chips were made of real potato, but the substance was so rocklike that I cracked a piece off a tooth.

"Suddenly a great horde of pale moths appeared, like helicopters, in the air. They are the precursors of rain, and always come before the monsoons. In a few moments vast drops came beating down from the black clouds above the dark blue mountains. At once the roads were like chocolate blancmange. We skidded dangerously. We saw a truck that was balancing amidst the branches of a tree twenty feet over the cud, hanging above a thousand feet drop.

"For the last part of the trip I lay at the back of our truck on the luggage. A tarpaulin had been put over us when the deluge began; so I was sheltered here from the elements and left with my thoughts. I lay in the darkness reviewing my life, thinking about my friends and conjuring up the past. It was like choosing old gramophone records to play again. Time meant less than usual, but in fact there were still many hours to be passed before reaching our destination. In spite of the tremendous bumping about I slept; only occasionally was I hit hard enough upon the head to recover full consciousness. When, at last, we had driven down through mountains, pinpricked with the native camp fires, past bullock carts travelling at a snail's pace to the music of their bells, their drivers asleep, it was half-past four in the morning. We had been travelling for a stretch of sixteen and a half hours, and were very cold."

By far the worst setback I have experienced in my long photographic career happened to me at this time. I handed over a package of about two hundred and fifty undeveloped films I had exposed, to be sent back by air for processing in Delhi. They never arrived at their destination. The airplane which took them did not crash; the package was merely "mislaid". Ceaseless, but nevertheless vain, attempts have been made to discover the precious parcel which, inadequately addressed, is still doubtless lying about in some

A casualty

Supply convoy climbing through Burma

After the battle, Arakan

waiting-room. A year has passed since then, and the chances are tragically small of my being able to fulfil my promise to my various sitters, to send them copies of their pictures. My apologies to the men living in jungle fox-holes, firing the twenty-five pound guns, to the Howitzer teams, to the Gurkhas of the 7 Regiment, to the men of the Queens Regiment, the West Yorks, and the others who showed such enthusiasm and co-operation. Myself I regret the loss of these pictures, for they were taken while I was in full possession of my first enthusiasm and energy, and when, it seemed, such exceptional opportunities presented themselves to my camera lens.

Whereas the scenery among the Chin Hills reminds one a little of parts of Scotland or California, the tablelands on the Arakan Front are unlike anything one has ever seen, except perhaps in the background of idylls and fantasies painted on Chinese fans, screens or porcelain. Around Maungdaw it is as if the compact mountain ranges have erupted and dotted the earth with hundreds of rugged hillocks. These hillocks are covered by Peepul trees, spreading their huge, dark leaves, and by bamboos, while the feathery undergrowth is pierced by long white shafts of pampas. The landscape is pastoral; so lush, sylvan and peaceful is the general aspect in the brilliant sun or moonlight that, in spite of the intermittent thuds of gunfire, one cannot quite believe that deadly warfare is being carried on nearby.

"We came unexpectedly upon a battle. During a picnic lunch in a ruined temple we heard gunfire. Where we climbed a flight of stone steps to discover what was happening, two over-life-size black-satin crows swooped down from the magnolia trees and carried off the remainder of our meal. So we moved on, down a disused road, through an overgrown village, once bombed, now abandoned and looking like the garden of the Sleeping Beauty, with exotic creeping plants sprawling over the half-destroyed "bashas" and summer pavilions, over the gutted motor-car still parked in its neat, cement garage. On again we went, past the farm where, in a courtyard, provisions were dumped — tins of bully-beef and packages of biscuits lay among hundreds of small eggs, gourds and the exotic vegetation of the tropics.

A group of young officers, with serious expressions on their sunburnt faces, were discussing the situation. During the night some Japs had come down through that jungle range there, and had taken up their former positions which, inadvertently we had not filled in before advancing farther. Now this enemy group, with a two-pounder gun previously captured from us, was dug into the earth snug as moles, and able to do quite a lot of damage to our rearguard. Several men had been killed, and the wounded at this moment were being brought back under fire. The stretchers were placed in the Red Cross ambulances, which the drivers manipulated on the rough roads with dexterity and compassion.

A young major appeared, his khaki battledress stained with dark, dry splashes of blood. "We thought you'd been killed," the others greeted him. "Are you all right? Better have your arm seen to, and if you can cross that bridge, do so quickly and on all fours."

Meanwhile, in the fields of paddy, Indian women accompanied by their naked children were still working, unmindful of the bursts of shrapnel. Bombing by air alone will send them seeking shelter.

At the outset of this particular war the Jap had already perfected the technique of jungle warfare. His ruses were wily, and at first we fell into some of his traps. He was continually popping up in unexpected places. Since then, the Jap has made mistakes. We are now accustomed to his oft-repeated devices and are more adept than he at adopting the tactics of feint and surprise. It is we who are now holding back our fire from his sorties, only to turn it on later with great effect for the real attack. We now never attempt to storm a hill under fire from flanking hills, but rather infiltrate by the "back door". Our troops are no longer apprehensive of the Jap. The myth has been exploded that as a soldier he is superhuman or inhuman. Admittedly, rather than be captured he will fight bitterly, for he knows that to be taken prisoner is to be without hope, to be written off as dead, never to return to his country. His orders are to kill himself rather than fall into enemy hands. But, it seems, life is sweet, even to a Jap; often those who are taken prisoner, knowing they have little to expect from their own people are willing to divulge secret information in the belief that they may be treated more favourably. Recently the Jap airmen have been equipped with parachutes, to be used only for baling out over their own terrain. There is a story of a Jap pilot being shot on his descent over our lines by an officious and outraged compatriot.

Here are a few more extracts from my diary: —

"It has been such a particularly lovely day to-day — the sky so blue, the sun bright, the air like crystal, that it was difficult to believe that as a result of these skirmishes tragedy would soon be visiting some families back at home. The magnolias are flowering; it is hard to remember that those two leaves gently falling from that bough have been snipped off by a shot from a Jap mortar getting into closer range.

" 'Quick, it's time to move off,' comes the warning. Yet the scenery makes it difficult to think of danger, or in terms other than those of holiday 'camping-out'. There, waiting in the shade of those mimosa trees, the mules, heavily laden, look as if they are carrying up the provisions for a large 'alfresco' party.

I did not realise this great packing-up of the company's kit after the preliminary moving forward was, in fact, a step in the progress of the war, that our lines were advancing. When the howitzers, camouflaged with every variety of branch, are fired and the whole hill quakes, it is as surprising as if hidden guns were fired among the rhododendron bushes at a garden party.

"On to a small town which a week ago we had wrested from the Japs. Previously it had been subjected to heavy bombing from both sides. Since the Japs took possession over a year ago no one has lived in the houses, which now droop dejectedly under a covering of tropical plants. The grass has grown everywhere — over the frontdoor steps, into the hall-way, into the shelters, pagodas and deserted temples. Strips of corrugated iron have been flung by bomb blast into the leafless trees and remain there, gesticulating like tortured souls. It is another Sleeping Beauty town, but there is no sleeping to be done here to-day. The Japs had bombed it again this morning, and when we arrived a battle was raging a few hundred yards away. We continued in an armoured car to watch the battle. An officer pointed — 'The Jap is hidden there in those bunkers. Although there are six of us to every Jap, still he sticks his ground with amazing tenacity. It takes time to winkle him out and kill him.'

"Among the paddy rice fields and the more open spaces the fighting has little of the aspect of modern warfare. The return to importance of cavalry, and the mules laden with ammunition, bring to the mind pictures of Stonewall Jackson and the Civil War. It is only when one sees the treatment of the wounded that one realises how conditions have improved. So impressed was the Army Commander by one forward hospital that he made the experiment of showing the troops the elaborate equipment that had been brought on mule-back from three hundred miles away. He was not certain that some of the men might not be alarmed, but the experiment had the desired effect. The troops were tremendously impressed; they saw the casualties receiving such careful attention that the unconscious terrors of being wounded were minimised.

"Some of the camouflage attempts are half-hearted: some tanks and armoured cars trundle along the open spaces, garlanded with dusty pinetree branches, looking like old Christmas decorations. Overhead the enormous kitehawks were wheeling high in the sky, and a series of small, black cloud-puffs appeared and disappeared. Suddenly one realised that, higher than the birds, some Jap aircraft were flying amid our anti-aircraft shell-bursts.

" 'You are instructed not to come out from cover to watch. If you wish to see the battle, your face must be hidden by leaves or camouflage net,' said the young officer, half-humorously. Our fighters were now in pursuit of the raiders, while, below, cattle were grazing and some white herons were hopping among the paddy fields.

"We called at a front line hospital. A lot of horrors. One man's face was contorted with pain as his multiple wounds were being dressed: a piece of shell had got him on the shin; this was particularly painful while the bandages were being taken off, but he suffered like a brave child. I felt rather weak, saying silly things like, 'It'll soon be better'. I daresay the psychology of the doctors is cleverer: they are quite rough. They rag their patients to their faces — 'See him now? Well, he's a jolly sight different from what he was two days ago! It's a pity I put back that tip of his nose! It was hanging over his mouth when he came in. Bloody funny!' And the victim laughs, genuinely amused. The doctor confided that this man had only survived through his guts and determination; that most others would have given up and died. One Gurkha had been kicked in the face by a mule; the result was appalling to victim and beholder. The Indians suffer stoicly. Some of the fellows in pyjamas with malaria said they would prefer not to be photographed, to wait until they were in battledress again. An ex-waiter from the Savoy begged me to go with him to photograph the grave of one of his friends, Corporal Silk, who rather than let his comrades be wounded by a grenade that had started to sizzle in the undergrowth, had lain upon it. The ex-waiter, after eighteen months of bully beef, looked very wan, had lost three stone in weight, could not keep down any food. The doctors were looking after him as best they could, but it was impossible to give him what he needed most — a special diet.

"The work of the front line doctors is one of the epics of the war. For instance there is Dr. Seagrave, almost continuously operating under fire. The old man's hand would tremble until it touched the flesh of his patient. Then he would slice the body open as if he were taking the rind off a cheese, delve into the entrails, scoop out the shrapnel, and start on the sewing up. That job finished, another would begin. A young man, who had been shot through the eyes, is brought in. 'No, he has been unlucky! He's just one that lowers the average. Too bad.' The old doctor shakes his head with a terrible look of anguish. It was as if he had never before seen such tragedy. Then the next case — a young man shot through the groin — the shrapnel goes in small, comes out enormous; a huge hole in the left side of the thigh — 'Ah, this scrotum wound's not so serious after all! He's lucky! Here's one of the lucky ones!'

"As we passed the river they were bringing in a corpse. A horrible swollen parcel. This strip of water is a godsend, as the wounded men can be transported by sampan back to a base hospital, without the agonising jolting over potholes that kills so many."

We drove in a jeep over rough roads which were being sprinkled with water by native women wearing dark, gloomy coloured draperies. This spraying, from gourd-like vases, seems futile, for it succeeds in laying the dust for only a few hours; but I am told that it helps to keep the roads from rapid and total deterioration. The rates of pay for this job are small, yet the women are like princesses doing their humble job with dignity and heartrending poise. Some of their features are wonderful. Their children help too, and throw water from old cigarette tins, jam jars or other little receptacles. While motoring over these craters we talked, most of the time, about subjects completely unrelated to the war, or our surroundings. The sun began to fade; the country became flatter. Everything became more sympathetic and feathery; the distance soft blue, the trees like spinach, and the humps of the hillocks dotted with lettuce-green undergrowth. The evening at once grew damp and cold, and we were thankful to arrive at a Transit Camp, where there were Bashas of bamboo, and — great luxury! — an orderly to attend to our every need.

"We opened a bottle of rum and were drinking in our tent when an eccentric old colonel appeared, and in a voice deep down in his throat, rasped: 'May I make my number with you?' My companion became truculent and asked in a surly voice: 'Why are you dressed as a colonel? You're not a colonel are you?'

"The old boy was rattled, and suggested he should go to another Basha. However he remained to amuse us and explained that he was an armament expert. He produced a large trunk full of weapons, like a property-box from a Ralph Lynn farce, also a black pei dog that he had bought for three chips. We eat an excellent dinner of well-cooked rations in a clean, light and congenial mess. Some of the smells here are wonderful — the charcoal fires, and the Indian savours of spice and cooking. After sleeping in fox-holes which are dark most of the day, and at night become almost rank, this fastness of bamboo seemed extraordinarily luxurious.

"We had been warned that there would be a great deal of gunfire in a few moments — 'It's us, not them.' Although I was told that the barrage was enough to split the eardrums, I was so tired after my day in the sun that I lay all night quite unconscious of the guns at close range, sleeping soundly and, thanks to the loan of the orderly's extra blanket, warmly."

January 19th — In the Jungle.

"I awoke early. It was not yet light. There were heavy drops on the bamboo leaves above my head. I pictured the rains making it impossible for us to leave this mountain peak: the mud is so soon churned up, and the narrow paths in the jungle become obliterated — traffic is at a standstill. However I discovered that the drops were caused by the heavy dew which falls all through the year.

"Even this strange life in the jungle has many similarities with civilian existence. Major Abbott, my cicerone, was feeling a bit 'piano' on awakening.

He sat on the edge of his bed delivering a soliloquy, while higher up the hill two young men were singing. 'Those half-witted fools, they wake up in the morning and talk such utter tripe to one another, they nearly drive me mad. I'm not liverish, but I'm too old to hear people sing at such an early hour.' Abbott, like many who hate to get up early, prolongs the agony by dawdling in bed long after he is called, and making the final decision to get out from the blankets when he is already late. Even so, he was quicker than I in dressing, for shaving is such a painful procedure and it takes me so long to nibble away with cold water. Abbott attacked his chin as if it were elephant hide. He scraped so hard, so boldly, that I thought his steel blade would split. The knife over the bristles made quite a loud noise. The dry shaving lather, under the lobes of the large hairy ears, escaped the perfunctory douche of cold water, and remained for the rest of the day.

"The Ngakyedauk Pass was closed while a convoy came over the mountain road, but we managed to extract a permit from the Military Police and took pictures of this extraordinary path that has recently been made through the hillside. Some of the going is still dangerous and, to warn people against falling over the precipices, some of the road is screened with strips of sacking, tied at intervals to dead trees. This makes a most curious and unreal effect with the elaborate panorama in the distance. That evening, in the setting sun, far beyond this honey-coloured foreground, lay a lovely landscape of grey hillocks and small hills with sharp formations stretching for many miles around — it was like an enormous panoramic background for a picture, possibly by Leonardo. Along this road, while the sun sinks behind the distant hills, the coolies hurry, barefooted, carrying enormous tree trunks for bridging, or long flexible poles with hanging baskets. The waddling walk of these people, paralytic and jellified, looks affected."

"We crossed the river in a sampan to visit a mountain battery. When we arrived it was late afternoon: many of the men were freshly shaved, their hair brilliantined: they were now relaxing and writing home — but they complained that there was nothing to write about. One man was having his hair cut. Private Sullivan was reading a novelette, "Sunshine after Rain". Young Pierce of Birmingham, with a mop of fair hair, looking like Ariel, was cleaning out a mug from which most of the enamel had been chipped. Others were now rereading the troops' newspaper, already enjoyed and thumbed by so many that its pages were limp and pulpy. They offered you a cigarette and then, if they liked you, brought out their wallet and showed you, for admiration, their most treasured possession — the photographs of their wife and family. They would tell you that they miss their mother's cooking, an easy chair to sit in and flowers arranged in a vase. The cooks were preparing the evening meal. Fred Ridden was straining the water from the cabbage. Most of the men eat their meals without any relish; so long as they can have a brew of thick char three times a day, they are satisfied. Many officers say they would rather have a cup of tea to stimulate them after a strenuous day's physical work than a tankard of beer. It seemed strange to find men and boys, from all over England, dumped on this extraordinary warren, climbing up and down the steep inclines, picnicking here for months on end, packing up and moving off to another hill."

"We drove in a jeep by the side of dried paddy fields from Bawli Bazaar to a place near Cox's Bazaar called Elephant's Point, situated on the long winding strip of sand stretching between the Indian Ocean and the jungle. Here at tremendous speed we raced across the hard sand in burning sun. It was a curious sensation. The sea was brilliant blue and silver. Here were health, sun, ozone — little wonder that I felt so well. Indeed I was now feeling so sturdy, contented and free from anxiety, that my whole appearance had changed. Even the vibrations I sent out were different. I felt years younger, and had learnt patience. We arrived at our destination — a holiday camp where service men, tired, rattled and in need of a rest, are posted for a week's

Jungle Recuperation Hostel

relaxation. The place delighted us. A group of a dozen Bashas and huge recreation buildings, all made of this honey-coloured bamboo, created an effect as pretty as anything I have seen in Florida, California, Mexico or the South of France. These buildings were erected in five weeks. Some of the B.O.R.S. on arriving here felt at first that everything was too grand for them; but they soon came to enjoy the taste with which the whole camp is run. After spending a few days of relaxation here, bathing, playing games on the sand, sleeping at night without having to listen for possible dangers, the men seem completely changed; eyes lose their look of strain: faces are sunburnt and lines are ironed out; and by the time their holiday is over they are anxious to be back with their regiment."

I left the fighting areas with many impressions jangling through my head. Much of the time had been spent in being uncomfortable and doing things that do not normally interest me; but I had been without anxieties. I had discovered that, degrading as this remote and primitive existence can be, there are compensations, and that even warfare may bring a feeling of physical serenity and peace of mind. After two weeks, my clothes were incredibly filthy; my face had entirely changed, both in colour and in expression. Even my hair was thicker, though only with dust . . . . Now I was heading for civilisation; the rut would be deep once more. In the jeep we raced towards my airplane and captivity. This visit to the war had been in many ways an escape from war. But at this time of the year the climate is ideal, and the weather suited for the picnic life that must be endured for weeks, months and years on end. I was able to sympathise with the R.A.F. officer who, on hearing that he was to be sent home, confided: "I'm beginning to be really frightened now. It may be such an anticlimax to go back to England, after all these years of thinking about the place and building it up in my imagination. I have glorified it all this time, and now that I'm really going back I'm afraid."

CHAPTER FIVE

Charivaria

I RETURNED to enjoy, with an added appreciation, the amenities and luxuries that are offered to the privileged in India. I will not try to describe my headaches, how I sweated in the summer heat until my clothes formed a sort of outer skin. I will not expatiate on the pleasures of taking a long bath, or hearing the clink of ice in a long drink, but will confine myself to a few "vignettes" collected during my tours of the next few weeks.

The scene is any Government House in India. In the throne-room, an assortment of respectable English and Indian citizens are assembled. The inevitable Belgian Consul and his wife stand next to the huge retired Colonel with high blood-pressure, who must avoid the brandy. One of the Indians wears a little shade across the width of his glasses, a most peculiar effect, as if he were wearing a Pullman-car reading-lamp. Some officers are in uniform; business tycoons, wearing baggy dinner-jackets of tropical weight, are accompanied by their scraggy wives. I feel sorry for the wives; they are married to men who are making great fortunes, but oh, they are paying for it dearly, looking thin and wizened, their complexions dried and wrinkled! In recompense they go to the races, they play bridge; and this dreary evening itself will provide material for future gossip. At the time they talk but little and take their pleasures seriously. Snobbery is their religion; their second Bible is the Warrant of Precedence. From the Governor-General and Viceroy of India (who is number one), each European of any standing has his number, his order of going in. Governors of Provinces are number two; the Commander-in-Chief, number four. If we descend the scale we find that Judges of the Federal Court have dropped to number thirteen, while Archbishops of the Roman Catholic Church rate fifteen (a). Small fry such as Baronets of England are marked down at twenty-three (a), Brigadiers at thirty-five, and a variety of officers such as Chief Electrical Engineers, Engineers-in-Chief of the Lighthouse Department, Superintendents of the Carriage and Wagon Department, at forty-seven. The Opium Agent Ghazipur warrants fifty. We have now sunk rather low in the scale, among the Electrical and Sanitary specialists, and, poor wretches, last of all, the Examiners of Questioned Documents.

There is no thought of arranging a formal dinner-table in any but the order of precedence. In small communities the same people find themselves, continuously and irrevocably, placed side by side, with nothing further to say to each other. It is not to be wondered at that the evening does not always go with a swing.

"Will you kindly form a line along there?" suggests a rosy A.D.C., with one arm and cursory manner. "Two rows please — come along now!" Some of the A.D.C.'s enjoy making the guests suffer. "They don't come to Government House for nothing," they say.

A long delay, long enough to make each guest fully realise what he is waiting for. At last a slight commotion is heard in the distance. "Their Excellencies," shouts the obstreperous young A.D.C.

Dinner is served on an enormous strip of table decked with bouganvillea. The inanimate faces of the heterogenous company are reflected in the row of silver cups, in which fronds of fern are placed in imitation of smylax.

Thirty servants, with scarlet turbans and bare feet, run around serving the inevitable banquet food. Each of the Governor's jokes is greeted with bursts of sycophantic laughter.

"Mercifully he seems in a good mood now," says A.D.C.2, sitting next to A.D.C.3 in starvation corner. "But I've seldom seen H.E. so rattled as he was this morning."

"Why?"

"On account of cotton and pins."

"Cotton and pins?"

"Yes, someone took away his graph of the Arakan Front, and I think I know who it was. But no one has owned up yet. How can these servants understand what all that mess is about? Anyhow, someone tidied it up."

"Her Excellency personifies graciousness itself, though she, too, had a bad morning. They put some flower garlands round her neck at the opening of the Agricultural Exhibition, and they dripped down a new dress she had had copied by the dzersi. She was so annoyed she didn't speak a word the whole way back in the car, and then, when she got back, she kicked up hell about the expenses of these dinners. Said we want quantity not quality, and we can't afford a shoulder."

The elegant Englishwoman next to me exclaims enthusiastically: "I *know* I must be *mad* as I've been *seven* years in India!"

On my other side, an elderly Indian gentleman tells me in a lifeless uninflected sing-song voice how he and his family are unable to obtain first quality rice. They have always been accustomed to it, but now they feel "uncomfortable" eating such poor stuff. It gives them indigestion; yet he, being a minister, cannot buy on the Black Market. He has always been a man of high principles, and in his position he cannot cheat.

The elegant lady turns to me again.

"Now do tell me how's dear old London? Very much changed? How are all the old places?"

"What sort of places?"

"Well now that leather shop in Grafton Street. Is that still going?"

"I'm afraid I'm rather vague."

The elegant lady alters her role. She now plays for sympathy:

"I've been away from England so long, I hardly know anybody there now. But when I go back I suppose everyone will say, 'Oh you were in India all the war?' and I'll have to answer, 'Yes'. You hardly know there is a war on here," she sighed. "You get irritated at not being able to buy a bottle of whiskey, and then you have to realise what other people are putting up with!"

From across the table a local business-man buttonholes me.

"I know your name. We all attain a degree of notoriety, don't we?"

Lady Blank, a living skeleton in wasp colours, has not addressed a word to her neighbour: she has eaten each course with a grudging determination. No sooner has the company risen and drunk the Governor's toast to the King Emperor, than Lady Blank shoots through the double doors, quick on the heels of Her Excellency. Lady Blank, after fifteen years of service in India, is not going to stand for that common little Lady Anon being placed above her. Her husband will be made to write to the Military Secretary for an explanation just as soon as they get home.

The company now retires to the illuminated garden and sits out in armchairs and on sofas, placed on Turkish carpets. It is a magical night; fronds of palm trees are silhouetted against the starlit sky. There is a crescent moon, and the Police Band plays "Merrie Englande" and "Poet and Peasant". The bandsmen are yellow and blue, with white spats.

At ten o'clock more of the European Colony are let in to the sacred precincts. A further display of Anglo-Indian fashions; some of the sailors of the

R.I.N., in immaculate white uniforms, are almost throttled by their high collars — Beaux Idéals of all novelettes.

An intellectual lady, in a taffeta picture dress with a berthe of old lace, leans forward:

"Isn't it extraordinary that so great a country as India should have fallen so low? There is nothing of promise to be found anywhere here to-day. No writer, no painter. The only hope for the young Indian is to go into politics; and the only hope, if the country is to regain vitality and honesty, is revolution. If Congress were to take over, they'd make the inevitable mess of it; the dishonesty and craft of the Congress leaders would soon be discovered — bloodshed and anarchy would follow — but out of that, some fresh life might spring."

An elderly industrialist leans forward and says: "India is a feminine country, all her faults are feminine ones," and he raises his glass gallantly.

A beautiful Indian in a pink sari says: "Whatever those faults may be, let us make them. Please allow us our own headaches. India for the Indians please."

The A.D.C.'s move everyone around, as if in a game of musical chairs.

Under a vast electric fan, like the propeller of an airplane, a lady in cornflower-blue lace welcomes a newcomer. "We were just saying that the problems of India only begin to get really confusing for someone after their first year here: then so many conflicting impressions arise, upsetting your balance, that you don't know where you are."

A young subaltern says: "I always say it takes a year to learn to hate India."

Two A.D.C.'s are standing apart, eyeing the guests. One holds a small printed card up to his mouth.

"H.E.'s already had the Sanitary Specialist's wife three minutes. It's time we got the Expert on Humus Heaps ready for him."

"Oh no, Mrs Bumface gets seven minutes, she's on Post-War Reconstruction, but look, Her Ex. is getting a bit browned off with the Brigadier, hurry up and take that old chap over, he's the Commissioner of Police, what's his name?

Sitting by a huge pot of cannas, a lady with blonde hair is telling an impressed group about her war work. She had organised the first ball here. It was called "The Glamorous Night"; it was wonderful getting so much money out of people. With a flick of a cyclamen chiffon handkerchief she explains:

"I can't tie bandages or nurse, but I'm super at organising dances: I'd like to become a professional hostess after the war, in London, and make bags of money."

At "The Glamorous Night" there had been "bags of people", but she got so ill organising these balls that her doctor had said: "You must stop your work, it's no use your losing an arm or a leg. You're very good at this, but you must take things quietly."

"But," she gave us a confidential wink, "he never knew that all the time I had a telephone under my bed and was calling up hundreds of people!"

The obstreperous A.D.C. has been relentless in chivying the guests from one seat to another. He has been determined that the party shall end as soon as possible, as he has a clandestine appointment down in the Grand Hotel Bar, which shuts at eleven o'clock.

Her Excellency is rather enjoying her talk with Lady Anon when the A.D.C. interrupts.

"I think, your Excellency, that His Excellency is preparing to say good night."

The guests are hurriedly thrown into line again. Their Excellencies smile with relief. It is the smile the dentist receives when his patient is freed.

"Good night! — good night! — good night!"

The cars are churning up the gravel, especially imported from England. But in the first limousine, leaving a wake of dust and small stones, is the rubicund A.D.C., mopping his brow and telling the chauffeur to drive "Jaldi! Jaldi!"

A young Chindit, whom I had met when I first arrived in Delhi, was now about to go to hospital suffering, after four months in the jungle, from jaundice, amoebic dysentery, malaria, and every other sort of tropical disease. He had had a beastly time. My friend said he was saddened that his men must continue fighting under such appalling conditions, and he himself felt that they were incapable of conquering Burma in this way. Some of the men were beginning to ask, "Who wants Burma anyway?"

He, like so many others, had only one desire, and that was to fight the war in Europe. He considers the Germans are the natural enemy; he said he did not think it "worth while to be killed by a Jap, but a German is fair cop! Each soldier fights for glory: there is no glory in Burma. The British troops will put up with anything, their powers of endurance are remarkable, but some of the men have been out here five years, and an old campaigner is a grumbler; there are many grumblers."

The casualties among his men had been nearly seventy-five per cent. His men had done a good job and had mown down relays of Japs who had attacked them with almost inhuman relentlessness. They counted thirteen hundred corpses around their barbed wire defences. When they had taken Magaung, only a week ago, he had come upon a hospital, abandoned by the Japs, who had left behind them thirty patients without food. As the English arrived, some of the patients fired from their beds; one threw a grenade, and others crouched around a grenade so that their stomachs were shot away. Finally, out of thirty men, only five were taken prisoner; the others had to be shot.

When later I visited the young Chindit in the Military Hospital he looked most pathetic as he lay, wearing green pyjamas, his face and body terribly shrunken by disease. His hands, coming out from flat paper-like sleeves, looked huge and predatory. He told me he had been having vivid dreams the last nights, re-living his experiences in the jungle.

"I'm not unnerved," he said. "Just depressed and tired. In so many ways it's been such a waste of time, and such a waste of the lives of so many friends."

I felt sad to leave him, for I felt doubtful if he would recover. A few weeks later he was dead.

The station was a vast dormitory. Many of the human bundles continued to sleep when the train came in with a deafening shriek. The carriages were so full that some of their would-be occupants hung, like tassels, from the doors and windows: yet everyone seemed to be enjoying the occasion, singing, laughing and spitting almost ceaselessly. The Indian is continuously washing out his mouth and washing his feet; and in many ways, he is more careful of his cleanliness than our own troops. But he thinks nothing of infection or of spreading germs by spitting. Cases of smallpox will be aired abroad, and corpses often left in a train. "Grandmother has died en route — just too bad." The remainder of the family go their way leaving Grandma behind to her own resources.

Kaleidoscopic views move slowly past the windows; in the late afternoon light everything looks its best. The vast plateau is piled with boulders, one on top of another, which are said to be the relics of a sunken ocean; labourers are going home in the apricot light of the setting sun.

In my compartment sat a scientist in cotton-research, with the thinnest torso I have ever seen. He spoke English with grammatical perfection, and explained that, but for the war, he would now be at Cambridge. I offered him part of my supper; he tactfully accepted a revolting ham sandwich. My provisions were very English and, though prettily wrapped in a banana leaf, showed little imagination. Later my companion's bearer came in with a luncheon basket full of the most delicious, highly spiced foods, exquisitely cooked. He offered me something that had a taste of fish but which was vegetable, something that looked like a fish and tasted of fruit. At each halt, no matter how remote the spot, the train was besieged by an army of beggars, sellers of food, fruits, tepid lemonade and coconut-milk.

My companion announced: "No doubt you've learnt that India is a small country, that there is one set that knows everything about 'everyone who is anyone'. If you visit one city they know all about it in another. I know you were ill in Calcutta and stayed three weeks in Delhi."

I wondered if perhaps he had been poring over a copy of *The Onlooker*, which shows pages of "Society snapshots", and describes how "charming in voile" Mrs. B — looked at the Bombay races on Saturday. Gossip is garnered from the leading towns under such inspired and inspiring headings as "Madras Musings — Poona Prattle — Nagpur Nonsense — In Lucknow Now — Calcutta Causerie — Peshawar Palaver — Bangalore Lore".

"Zita", in the fifth year of war, writes an article on "The Art of Wearing Jewellery". Here are a few gems of advice: —

> "The peeress wears hers with perfect poise. Don't proceed to don mechanically all the diamond and ruby jewels that you possess, instead choose one outstanding piece from among them, or better still, a couple of emerald ornaments; this will provide colour contrast and suggest individuality."

and again,

> "So many women take their jewels as a matter of course. It is taken for granted that the best diamonds must shine at the biggest parties, and there the matter ends. Now don't go to the other extreme and avoid jewellery altogether. A woman has to be particularly lovely or particularly chic to do without jewellery for evening wear."

A friend of mine went one better than "Zita". She wrote: "Don't bring any jewels with you out here, the elephants have much better."

Hyderabad: The birds fly into this house as if it were an aviary; at night the insects create a fog around the electric light, bats rush in and out unremarked. Two small birds managed to make a great amount of mess upon my shaving tackle and over the wash-basin. One is accustomed to ants hurrying over everything, but I was startled to find two frogs in the bath.

From the heights of the Chars Minars, in the centre of the town, one sees best the broadened thoroughfares and the many pretty houses with elaborate windows carved like shells, and lacework verandahs. No one particular street is devoted to brass or spices. Unlike most Indian cities the streets are extraordinarily clean and sweet smelling, and the various kinds of shops are dispersed, so that one finds a Gun and Ammunition store (yes, anyone is allowed to carry arms) next to a nut shop, which is, in turn, neighbour to a flower shop (no flower remains long on its stalk, but is soon beheaded to become part of a garland). Only in the Moti Market, with its extraordinary, white, pear-shaped columns, are all the birds brought together in tall, pointed, cane cages.

The coffins of former Nizams are covered with scarlet cloth, which has faded to become a sad and beautiful pink, and strewn with half dead jasmine blossoms.

The temples at Ajanta, of which the remaining caves, halls and monasteries are India's proudest artistic possession, date from nearly three hundred B.C. to about six hundred A.D., when Buddhism was expelled from India. Although they have been covered with silt from about seven hundred A.D. until their comparatively recent discovery, much of their rich carving remains, and the pigment of the wall painting with which they were decorated is still astonishingly vivid. But not only their extreme antiquity makes these paintings remarkable. The stylisation of the human figures is interesting; the decoration of ceilings, flowered and bebirded, is delightful, in the manner of sixteenth-century Italian villa decorations; and it is fascinating to study the detail. We can see so much of the life of the times; the shops, the processions, the reception of a Persian Embassy by a Rajah, a snake-charming scene, and a hunt, with antelope and hounds. We can enjoy the precision of the paintings of jewellery, and of ants climbing a tree. But æsthetically I feel these paintings, which we have come to know so intimately from reproduction, have been over-praised. The drawing is often weak, the faces too pretty, the fingers too pointed.

To me, the temples carved out of the rock at Ellora are infinitely more extraordinary. A stupendous tribute to religious emotions, they were created by medieval priests, who were at work over a period of six hundred years; they are perhaps the world's greatest work of devotion. The shrines are shaped out of the rock and carved downwards in imitation of wooden buildings; the exactitude of the proportions of the columns, and the expressive vitality of the stylised figures, all carried out on a gargantuan scale, are one of the most eloquent relics of India's greatness in the tenth century. There are twelve Buddhist caves, fifteen Brahman and five Jain temples. The rows of Buddhas forming colonnades, Buddha in every pose, preaching and in meditation; the figures of Siva dancing, sleeping, in the lotus position; the carved angels; the flying nymphs, so strong and expressive in movement — are not only the finest remains of antiquity in India, but are among the world's greatest works of art.

The Kaibasa, the abode of Siva, is carved entirely out of the living rock, with a courtyard, two gigantic stone elephants, a screen, a shrine two storeys high, and beyond a temple containing a hall over fifty feet square, borne by sixteen wonderful columns on a base carved with a procession of elephants.

In the strong Indian sunlight, the highlights on this rich carving are as dazzling as silver.

We drove into the mountains. A shoot had been arranged — my first shoot. I had never realised before how simple, to the point of foolishness, these expeditions can be. I had imagined a certain amount of risk was involved; but not on this occasion at any rate.

It appears that the preparations for this outing had started here last night, when a goat was supplied for the consumption of a female leopard, who had been shot at her dinner. The male leopard had later so enjoyed the partly eaten meal that, doubtless, he would come back to-night at eight o'clock sharp. Another wretched goat, provided as bait, was lead out and staked. The sportsmen retired behind the foliage-covered windows of a small concrete building, strategically placed ten yards away. Here we took up our positions to watch the misery of the goat.

Left to its own resources it became apprehensive. It felt lonely on the end of a chain, with the mountain landscape fading into darkness around it. Plaintive bleats rang through the canyons. It walked around its stake in circles. It became panicky one minute, its head darting this way and that. Then it seemed merely miserable, its front legs collapsed, and it lay still whimpering. But, in abject terror at some noise, it was up in a flash; again its head moved to and fro, listening. I shall never forget that blunt, rock-like profile , seen through the black leaves of my window. I wondered why I was allowing myself to be party to something I considered so unfair and ignominious. It was awful to see an animal suffering the mental torments that, mercifully, one has left behind one in the night-nursery. Yet I must admit that I had worked up a certain blood-lust. It was exciting to watch and wait in the silence for the great moment when the lurking leopard would spring out of the blackness of night, and the loaded gun go off. Though we realised there was little chance of the leopard escaping (a miracle if the goat were not killed), our eyes were popping out of our heads with excitement. Nor was there a chance of the leopard getting at us in our concrete hiding place.

So we waited. But we were out of luck, or perhaps the goat was in luck. After one and a half hours no leopard came, and the Colonel who was organising the shoot suggested that it was useless to wait longer. The anticlimax was crushing, the fatigue terrific. Even the goat had become too tired to bleat; it delivered itself of pathetic little grating croaks, like whimpers of despair, and lay down to sleep. Its joy when we came out of our hiding-place to unleash it from the stake was worth any disappointment that murder had not been committed. The goat had, at any rate, one more night to live.

Married couples fight among themselves; the climate makes them crochety. The husband says, "We will go this way."

"No dearest, this way," replies his wife.

"Please don't contradict me, I was trying to tell Beaton that the Buddhist art from India hurtled out to Java and China, and ricocheted back here much more Chinese than Indian; that is why you see so much Chinese stuff in India. But of course, if you will interrupt me . . . ."

The nagging continues, often at the expense of the servants. After dragging my weary bones around various courtyards, the stables, the swimming pool, the rock garden and tennis courts, I would long to sink into a chair; but I am taken to see the pets' cemetery and, as always happens when I arrive in a new place, I am pumped to the gills with information that I cannot possibly imbibe. I try to concentrate, then my mind wanders. It is extremely good of people to give me of their precious time; but invariably an argument springs up completely above my head.

"No, not in the fifteenth century, in the thirteenth."

"Not Deccan, Swatt."

"Now you must go to Benares."

"Stuff and nonsense, Benares is a complete waste of his time; a nasty, dirty place full of smells."

The mother hands me a silver frame. "This is a photograph of my son. He's just been asked to contribute to a magazine, I can't remember its name."

The father volunteers the information: "It's a provincial magazine of some sort."

The mother smarts. "London!" she contradicts, and doing so, she kills her mate with a look. The father comes to life again and retorts: "Provincial".

With fury in her heart the mother reiterates, "Excuse me it's a London magazine — I have got the letter — my son wrote the letter to me, and it's a London magazine."

"Provincial," says the father sullenly.

Suddenly the wife calls "Boy!" A very old man runs in with a turban on his head.

My hostess has lived in India for many years, but she does not speak the language.

"Turn on the fans."

The old boy goes out and turns off the electric light. We are left in total darkness. A major operation is launched before it is possible to get the "boy" back to turn on the lights again, and then to get the fans going.

My diaphragm became constricted. I felt feverish. By the end of the morning I realised I was in for some sort of 'flu fever. After lunch I had to take to my bed. I sweated, swallowed aspirins, sweated more, and the doctor came. He was amused at my taking six aspirins, and showed rows of teeth. He would come again. I didn't feel very ill, but more comfortable in bed than up; but I had a high temperature. Later I sank lower into the bed; my body ached a good deal. I wanted to have the pain broken up by massage, but this was not possible. By degrees I accepted the fact that I had no fight left. I rather enjoyed being a victim of some tropical disease. I quite willingly relapsed into a complete state of invalidism. I felt as if every bone in my body were broken. For days my body ached in the most unexpected places — on the shoulders, in the small of the back, behind the knees.

I recovered. Then I became much worse. One night particularly I thought would never end. Soon the fever worked itself into a crescendo of pain, sweating and general misery. The doctor told me that this was the normal course of Dengue Fever, which I must have caught from a mosquito that lives by the sea. After the very bad night I would recover, the doctor said. I did.

We motored along the roads fringed by Casuarina trees. Women, wearing draperies that gave them the appearance of fluid Greek statues, were working among the sugar canes, the wheat and brilliant green grass like clover on which the cattle live, which yields a harvest almost all the year round. One of the women wore a faded scarlet rag thrown over the head and body — a most unpractical costume for her job of threshing or gathering wood, but more beautiful than any other. In these parts red seems to be the fashionable colour, ranging from burgundy to squashed strawberry. The natives, in their bullock carts, wore crimson turbans of enormous size. The bullocks, placid and beautiful, wore appealing and pathetic expressions on their faces, and some had white dewlaps, like fish fins. Their horns flew out in huge, baroque volutes.

The village of Purana Ghat, of colonnaded, pale yellow buildings, of columns and pagodas of thrush egg blue, of massive archways under which the bullocks, harnessed to their heavily loaded carts, shelter from the rains, leads to the town of Jaipur. Of all the cities of India I have seen, this is the nearest to the dazzling pictures of one's childhood imagination.

Laid out in the eighteenth century, the wide streets run parallel. With its open squares and pleasure gardens, the city is so well planned and spacious, and its colours so harmonious that the general effect is of an almost dream-like leisure and serenity. Thanks to the rule that all houses must be painted a uniform coral colour and façades embellished with birds, fish and flowers painted in white, the effect is that of a Moghul miniature brought to life. The populace sport particularly bright coats or turbans; and when they bring out to dry in the sun billowing, freshly dyed lengths of vivid yellow or magenta muslin, one feels that nowhere else in the world have robust colours been used to produce such refined and subtle combinations.

Small carriages with hoods of the eighteenth century, shaped like pagodas or like sedan chairs on wheels, disgorge Rajputana ladies and their children, with heavily painted eyes, all wearing a mass of jewels, and crimson and dark cherry-coloured draperies. The dark faces of young men wearing peppermint pink turbans are seen peeping inquisitively from the balconies of carved white marble that are inset with coloured glass in floral designs.

In the inner court of the Zenana Palace, women in daffodil yellow, apricot and orange draperies polish the white marble columns. A young man, wearing a pea-green turban and a lilac coat, spends his morning lolling against an archway and looking like a figure painted on enamel five hundred years ago.

The legacy of this wonderful city has fallen into safe hands. The Maharajah of Jaipur is a young man with a proud appreciation of the beauties of his state, and a keen interest in building anew. The Rambagh Palace, with its ivory lacework, is as near to the ideal Indian Prince's Palace as any modern structure can be. Elaborate building schemes have been held up during wartime, but the Maharajah intends that his state shall become one of the show-places of the world. When this war is over, and travellers are able once more to take to the air for pleasure, Jaipur will become a great centre. Visitors will be well looked after in State hostels. They will not have far to go for sightseeing, for every vista provides a surprise. Their chief goal will be the Palace of Amber, perched in the hills that cup this lovely town. Apparently carved of the rock from which it soars, this castle is of such delicate proportions that it seems to have been built on stilts: Avignon comes to mind. The formal water-gardens and parterres are French in character; but the corridors with thick walls, cool vaulted rooms with windows placed to catch the evening zephyrs, and the terraces for moonlit supper-parties, all show the hand of the Indian architect. The interior decoration of the Palace is of a marvellous intricacy and richness. Some of the rooms are ornamented with mirror and plaster filigree, and when the windows are shut and the candles lit, they seem to have been invaded by a million fire-flies.

The Maharajah is particularly fortunate in his Prime Minister, Sir Mirza Ismail, a man of Persian extraction who combines energy and business sense with a poetic love of beauty. He is the arch-enemy of corrugated iron, brass bands and almost everything else that is crude and vulgar; one can be sure that, under his guidance, there will be no mistakes in taste, and that Jaipur will retain its original charm without ever degenerating into a mere museum-piece.

Besides supervising the architecture of the city, Sir Mirza Ismail also acts as Sanitary Officer. Jaipur to-day is so well organised that it has an air of almost Dutch cleanliness.

It is a thrilling experience to motor through the streets with the Prime Minister, accompanied by his secretary.

"That water over there overflows from the hills — see to it that there are more drainways." "Get rid of that black border." "All those houses must come down." "These streets must be enlarged by five feet, make a note of it." His car moves forward and he points through the window, gives his instructions and under his administrative authority, the miracles are set in motion. Already he has achieved in a short time remarkable metamorphoses. His plans are as numerous as his inspirations.

Over the mountains and down through the gorge, Sir Mirza has restored to beauty the temples of Gulpha, where a spring (supposedly a reappearance of the Ganges), rises, to which pilgrims come in their thousands for libations. The temples have gained a new life, their façades painted afresh with colours as appetising as those of angelica or preserved cherries. Hundreds of monkeys, clustering together on the ledges of the temples, watch the turbaned artists renovating the frescoes. A man with a strong face and long thin legs, wearing only a dark brown drapery falling from a white ring around his head, jumps across the rocks, with a flat basket, to feed the monkeys. He knows each of them well: he is determined which is to receive his allotted portion. Sometimes, just as one animal behind him is about to leap for a forbidden crust of brown bread, he lets out unexpectedly with a backward movement of a leg.

Inside the temple, the musicians, with painted forehead markings, wearing poisonous colours, sit by the hour making strange sounds on peculiar shaped instruments. The temple assistants, dressed from head to foot in blood scarlet, hurry past with plates of steaming food, curries and soups; hang garlands of marigolds and roses round the shoulders of visitors, or sit vaguely watching the acrobatic contortions of some of the worshippers paying obeisance to the glittering Goddess.

The cook had prepared for us a concoction of milk of almonds, rosewater, carminum nuts and eight ingredients of which Hashish, or Bhang, was the principal.

One of the effects of Bhang is that it makes everything appear humorous. Another is that strange things happen to one's sense of time.

The drink was delicious. We sat on cushions on the floor, expectantly. We waited for results, and were disappointed. So we had dinner. Still no effect, and our laughter was slightly self-conscious. This was thoroughly disappointing.

We had finished the brew, and there was no possibility of getting any more.

My friend started to giggle a little, but I was not amused. If other people's amusement is disproportionate one feels suddenly sobered. A little later I noticed that my hands felt soft and boneless, the skin unusually silken; I could not quite feel the extremities of my body. But the evening was a disappointment — too bad!

We sat in the drawing-room prepared to spend the rest of the evening in conversation. We had not been sitting there for more than a moment when the drug worked with such tremendous force that, for the rest of the evening, we were dazed. The experience was not altogether pleasant: for, although this is a laughing drug, and I laughed inordinately, the sensations were so far beyond my control that I was rather anxious and apprehensive as to how the evening would end. My friend, as he sat regarding me with pink blotched face and pearly grin, seemed to look quite different from any former picture I had had of his appearance. Outside, a strange metallic clicking sound was heard. It struck me as humorous that it should continue so long, and, imagining it to come from a bird, I remarked upon the bird's insistence. I realised in a flash that I was mistaken; but my friend's surprise, and his incredulity as he repeated "a bird?", seemed to me so terribly funny that I laughed until I was unconscious. In my chest I had a feeling of strange constriction; I continued laughing when I re-emerged into semi-consciousness, and my companion laughed at my amusement. I tried feebly to tell him that he looked different, that he reminded me of a school friend named Dudley Scholte, but I was tongue tied.

For the next two and a half hours we were violently drugged. The hallucination seemed to last an eternity. A reaction that would normally take only a fraction of a second seemed now to continue for ever. One wondered if the other person could read these long deliberations that were going on in one's mind. I noticed that my friend was far from reality, and sat smiling in utmost intoxication. The proportions of the room had changed. The distance across the floor suddenly seemed as large as the Atlantic; one's eyes could hardly travel to the far end of the music-room. The arches under which my friend sat assumed cathedral proportions, though, in fact, they were only a little over six feet high.

So submerged in intoxication were we, that I wondered what would happen if some visitors from the outside world were to arrive. We were incapable of speaking consecutively; the effort was too great, so that one gave up with a confidential look. Neither did one know if one had voiced a remark, or whether the thought had been so vivid that one merely imagined it to have been spoken. Somehow, one felt that speech would break the spell, and one did not wish to break it. But the spell continued and, occasionally, a reply would prove that one's attempts to hang on to sanity were not in vain. But my friend's brain was acting better than mine. I was able to understand him to say that this drug gave a far greater degree of intoxication than any drink. I could not have managed such a sentence. He continued, "Already we would have passed out or been sick, with an appalling hangover."

I heaved with laughter, and with tears coursing down my cheeks muttered, "We are beyond speech," and again was convulsed.

But although I was "in extremis," I was anxious not to forget this experience, and kept asking, "How much of this will we remember?"

My friend had a particularly beautiful and elaborate gramophone, with loud-speaker relayed from the ceiling: we decided to listen to some music. I have never appreciated or understood music so clearly as I did then: each instrument in a large orchestra was heard individually, with extraordinary distinctness. I was able to follow the construction of a piece as never before. An Indian song was played — some Spanish music — a Russian march — and then some very hackneyed Debussy. I could not concentrate for the entire length of each piece, but when I surfaced to enjoy a few bars, their beauty was astonishing.

I remember that we went out of the room to look at the servants playing high jinks by the garden door. To move across the room was a great feat. We watched two houseboys putting up the mosquito nets over a bed, and they too giggled hysterically. Most of the household had taken a few sips of the potion, and there were giggles at the back of every door.

Later I paced the room attitudinising, experimenting and trying to find out how far one could see, how much it was possible to regain consciousness by exercising the mind. My friend watched me in a trance. Not only was perspective altered, but the stereoscopic values were those of a badly equipped peepshow in a museum. My friend's body was flat, as if pasted on cardboard. There was a great distance between him and the table, and another vast jump from the table to the curtain. One's time-sense had broken down with extraordinary results — ten minutes would pass in a moment: a split-second would seem like many hours. The journey to my bed seemed to take an aeon.

Next morning I awoke with no headache, but feeling placid and unconscious of my body — this was an advantage later when sight-seeing. For days I was slightly under the influence of the drug, and laughed easily. I was interested and pleased to find how much of the experience I could remember, and how my appreciation of the music had been a real and lasting gain.

Benares, the sacred city of the Hindu World, was looking its most squalid in the mire that had accumulated after a week's rainfall. Along the edges of the Ganges, that holy but muddy river, people were washing themselves in the picturesque but unhygienic manner of which Miss Mayo has given us so hilarious a description. Yes, there were the corpses wrapped in cotton, in the centre of a pyre producing blue smoke. Yes, there were the scrofulous pei dogs, there the sacred cows nosing about among the mess, there the oozing walls, the indecent carvings and the temples redolent of butchers' shops, with their shiny tiles and smell of meat. There were the old marigold petals strewn among the muck.

This is India without neon lights, cinemas, cabarets or American standards of plumbing. This is Hindu India, almost unchanged, in spite of its sacking by the Mussulmans and the passage of centuries.

I stayed as the guest of a young Swiss, who came here for a short holiday seven years ago, and has remained ever since. He lives as an Indian, has adopted the Hindu religion, and in wartime manages, remote from the chaos of Europe, to continue his studies of Indian music and philosophy.

In this household there is no hurry, no dressing-bells, and meals are served whenever anyone feels hungry. I felt that life here had much of the truthfulness which is essential to the happiness of any artist.

My bedroom consisted of a bed on the terrace of the Zenana's wing, which I shared with a delightful creature, the Sacred Cow. Strangely enough the women's quarters, painted terra-cotta red, were uncompromisingly dour and masculine in effect.

I washed in a series of brass bowls. There were no easy chairs. We sat on the floor, leaving our shoes on the threshold. A man came to sell paintings, some of the eighteenth-century, some obscene. We eat delicious foods with scented flavours and aromatic spices, sweets of orange jelly and rose-leaf — as much a perfume as a sweet.

What would we do? Have some music? The musicians arrived. I appreciate many of the rhythms and the intricacies of some of the variations, but I cannot follow the course of a piece of Indian music; nor, having chosen a special scale, as any appreciative listener does, do I know what notes must eventually be struck. But the sounds that issue from some of these fruit-shaped instruments are soothing and soporific.

My host tried to give some account of the advantages of the Hindu religion. The interpretation by the old gurus of the Holy Book was thrilling. Was it not foolish to challenge their superior knowledge? Why be incredulous if some unintelligent person were to state: "This electric iron is cold, but if you plug this wire into the wall it will become hot!"

"So," said my friend, "why be sceptical if he says that from the worship of the cow one can receive an amazing impression of divinity." It cannot be denied that my host possessed a rare calm and serenity.

By now I had travelled many thousands of miles in India, and had seen, in a few months, more than many whose lives are spent pegged to some one spot on this gigantic peninsula. I had seen displays of surpassing wealth and acute poverty. I had discovered that in this country where eighteen million people make their livelihood from the land, and where the rapid increase of population creates troubles on an ever-mounting scale, one great storm can bring to millions famine or bounty. I had never agreed that India was a "first-rate country for second-rate people." England has been wise in sending forth from her shores, as administrators, only the best of her race, for the Indian's perception of character amounts to an extra sense. Against India's historical background of tyrannies, corruption and cruelty, of internecine wars, famines and other decimating calamities, the efficient work of five generations of English service must command our respect. British engineers have developed huge barrage systems, bringing to fertility immense tracts of desert; canals and railways have been built across the country. The rise of Indian industrialism has been rapid. The Government Services, political, medical, police, judicial and educational, have been run with efficiency. The British have shown aptitude for careful administration, and have learnt that by one peremptory order they could not overthrow the tradition of thousands of years; that man's nature is subject to circumstances, differences of climate, food, soil, education and religion. The British have shown respect for the deeply religious feelings of the Indian. On this point let me quote the Abbé Dubois, a medical and ecclesiastical missionary, remarkably free from theological prejudices, who wrote at the end of the eighteenth century an extraordinary work on "Hindu Manners, Customs and Ceremonies."

> "Accordingly there is not one of their ancient usages, not one of their observances, which has not some religious principle or object attached to it. Everything, indeed, is governed by superstition and has religion for its motive. The style of greeting, the mode of dressing, the cut of clothes, the shape of ornaments, and their manner of adjustment, the various details of the toilet, the architecture of the houses, the corners where the hearth is placed and where the cooking pots must stand, the manner of going to bed, and of sleeping, the forms of civility and politeness that must be observed; all these are severely regulated. Nothing is left to chance. Everything is laid down by rule, the foundation of all their customs is, purely and simply, religion. It is for this reason that the Hindus hold all their customs and usages to be inviolable, for being essentially religious, they consider them as sacred as religion itself."

Yet in spite of almost unsurmountable difficulties, improvements have been made in many matters of hygiene. Widows and lepers are no longer buried alive. Yet, as the late Robert Byron wrote in his brilliant "Essay on India":

> "The effort of the East, in civilisation, has been primarily metaphysical, that of the West social; Western man's betterment has been achieved through political experiment, in the East, concentration or discovery of good by thought and ecstasy."

Perhaps it is owing to its climate that the English have never colonised India. The majority of Englishmen who arrive to give their services to India think only of the day when, their task finished, they will return to their native land. Yet the Insurance Companies work on the calculation that these men will survive retirement in England only three years.

I have come across dreary examples of the intolerance and lack of imagination of a few British working in minor capacities, just as I have become exasperated with the weakness of character and idleness, apathy and desire to shelve responsibility of certain Indian officials. Much of the criticism we have heard of existing conditions in India comes from people who, liking to see all nations of the world living according to exactly the same standards, mistrust, or even consider as uncivilised, everything which differs from their own creed.

In the great heat of the summer I had travelled by electric train to the mountains, and enjoyed the perfumes, so fresh and welcome, of the damp moss, ferns and palms. In the Indian Hill Stations, Scottish Hydros spawn among a vegetation which seems to belong more to a Victorian Britain than to the Himalayas. Virginia creepers circle around the ironwork verandahs, and one is back among the faded snapshot albums of croquet parties and groups on the porch steps.

I had formed many quick and easy friendships; I had come across some remarkable people; one of the most intelligent men I had met was a young Indian poet. I had begun to rely upon the thoughtfulness of my bearer, who tended me like a parent during my bouts of fever, and the note of whose character was struck by Robert Byron:

> "That innate sense of propriety, decorum, acceptance of stations, which is a result of his profound conviction of human inegality. He has no desire for equality, knows no hatred for the rich and propertied, in fact is against his most deeply engrained instincts."

I had been impressed by the old doctor, the only man in India who can do the eye operation for cataract, who brings back the eyesight to two thousand Indians during the six weeks when he leaves the hospital where, for the remainder of the year, he carries out his experiments with American doctors training under him. It would take only one day's illness, and a short glimpse of their work, to realise that the over-life-size matrons in charge of large hospitals are the salt of the earth. In England there is one nurse for every two hundred people; here one among thirty thousand. These matrons are busy teaching Indian women to nurse, their operating theatres are filled with Indians, Indian anaesthetists, surgeons and nurses; yet continually they feel an undercurrent of resentment and discord from those in their charge, and find notes placed on their chair or desk with the message to "Quit India". The Provincial Governors allow poor funds for their work, and their own salaries are pitiably small. One of them told me that her great pleasure was to go and look at the Asia Crafts Shop, at the wooden bowls which were "so full of colour. They don't mind at all, if you don't buy, if you just look around, and occasionally I *do* buy something — but that's for a wedding present."

I had been impressed by the red carpets at Government House, and had received much enjoyment quietly watching behind the scenes the A. D. C.'s at their various jobs, and listening to the young subaltern, with one arm and a large moustache on the telephone, saying "Her Excellency says let it simmer till seven."

Many unforgettable pictures had been flashed on my mind. The Rajputana women, like goddesses, in red, carrying the red bricks on their heads with which the American barracks are being built at Delhi: the highly-coloured groups in their fluttering scarves drawing water at the wells as they have done for the last two thousand years: the three priests hurrying, in pale apricot draperies, between the trams in Calcutta, in this paradoxical setting looking like Tanagra figures. On the rocks splashed by the waves of the Arabian Ocean, to the sounds of an old man beating silver discs, the Indian dancers postured, grimaced and went through the formal rituals of the incantation dances. A coolie, with a huge sheet loosely wound around his head, like Aladdin's genie, went before me with a lamp, revealing the various treasures from the dark past in the painted caves of Ajanta. In the factories and mills occasionally one would come across a young person occupied at some humble task, working out his or her god-appointed destiny with such remarkable dignity and grandeur that one felt near tears: and often the conjunction of youth and physical perfection created an effect that was more akin to magic than to human matter.

I had been awe-inspired by the Indian women, married in their teens, mothers of many children before reaching the age of twenty, working so hard amid grinding poverty, carrying heavy, gourdlike pots of water from the well, cooking, taking their husbands' food to the fields, looking after the children and animals and even remaking, with cakes of cow-dung, the walls of their homes. All these tasks they do with a grace of movement that has the dignity of an empress.

I had discovered how detrimental to the brain the Indian climate can be. I had telephoned to Natarajan and said, "I want you to do three things for me." I enumerated the three things. Natarajan replied, "I must put those down now before I forget — one, yes — two, yes — now what was the third thing you wanted?" Neither he nor I were able to remember.

I had seen an electric light bulb burning palely in the sun; no one could be bothered to turn it out — yet there was a scarcity of electric bulbs. I remember thinking, "This is typical of India."

In spite of a wish to trespass into the elusive company of the inhabitants of this country, I felt I had seen little of the real India. I had not more than begun to experience a few of her many sorts of weather and geographical conditions, that range from the perpetual snows of the mountains to the stretches of scrub where the heat rises to one hundred and thirty degrees. Yet, now I must leave.

I ran into the house and did rough packing. Within three minutes I was ready for departure. I distributed largesse, as the expression is, to each servant. But the staff awaited the moment for me to step into the car, before running out with a large box marked SERVANTS BOX.

"SERVANTS BOX!" — "SERVANTS BOX!" they each took up the cry, and, before my eyes, pushed the rupee notes into the slot; each one laughing with childish amusement, "SERVANTS BOX!" — "SERVANTS BOX!"

As I drove away I was cheered by the household — the sweeper, the cook, the laundryman and the kitmagars, all shouting and laughing, "SERVANTS BOX! — SERVANTS BOX! — SERVANTS BOX!"

Ashcombe, 1945

Indian gunners

Indian Album

To

His Excellency Field Marshal the Right Hon.

THE VISCOUNT WAVELL

P.C., G.C.B., G.M.S.I., C.M.I.E., C.M.G., M.C.

Viceroy and Governor–General of India

This book is dedicated

Foreword

One evening, when I was photographing scenes of typical life in Calcutta, I had stopped on my way home across the Maidan to take another picture, when an elderly Scot came up to me. "Young man," he said, "you're going to be disappointed — you're making a very great mistake taking a snap shot into the sun. I've been in India thirty-five years, and have learnt that you can't do that. Only if you go up to Darjeeling, and you get out of bed at five o'clock in the morning, can you get the effect you want." I remonstrated. "No young man — let me warn you, you're going to be disappointed. Photography in India is a tricky business!"

But from the moment our seaplane glided down on to the Sacred Lake of Rajsammand, and I received my first glimpses of India, I had known that I would not be disappointed. The glittering white palaces of Udaipur towered in the distances: a small boy, in white and scarlet, wearing an enormous turban, came down the flower potted terraces with a platter of highly coloured refreshments. I longed to start, then and there, on my photographic mission. A few hours later, in the spiced and clover scented streets of Gwalior, where the wooden houses have their balconies, shutters and pediments elaborately ornamented with flaking paint, and everywhere there is a mass of draped pedestrians, overflowing carts and pampered sacred animals — I knew that photographing in India would be an endless pleasure.

These first glimpses were but the prelude to a succession of pictures of such extraordinary beauty that I was never able to "take for granted" the visual aspect of the country, and my camera was able to catch only a fragment of all that I saw. Yet among the collection of negatives I have brought back there is a large range of subjects: the factories where precision instruments for aircraft are made, the gun and rifle workshops, the social institutions, the Royal Indian Navy, and all the varying aspects of India's effort in the war; idyllic village scenes around the large pool, the focal point of local life to which shepherds bring their flocks to drink and a horseman gallops on a white steed. There are the pearly mosques, the sculptured temples, the riotous gardens, the rural schools and the great universities. Perhaps the post-graduates of the Science School would be giving pressure to liquids by drawing coloured water into their mouths, or perhaps making experiments that would end in a minor explosion and much laughter; the farmers would be ploughing with agricultural implements that are of a Biblical aspect, and the women in the villages, in mustard and red draperies, would be drawing water at the well as their ancestors did, three thousand years before. In the jute factories the women, heavily jewelled, would be lining up like caryatides, with the sacks of jute carried on their proud heads. In a remote village, outside a strange building — a modern version of an Italian Renaissance palace, with cherubs garlanded with fruit and flowers — I remember the beauty of the spectacle of the festival that was being celebrated. Many goats had been sacrificed; now the crowds were busy shopping, buying bread-fruit, hairclips and earrings. The women seemed like lotus flowers; all wore slicked hair above their neat flat features, and the colour schemes contributed by their brilliant saris were always unexpected and dazzling.

The extraordinary natural dignity of the natives of India was something by which I was continually impressed. I shall always remember the young widow with her five-year-old son, who had been fetched from her village to be the centre of admiration of a great concourse assembled under the rose-coloured walls of the Fort in Old Delhi, while the Viceroy gave her the Victoria Cross, posthumously awarded to her husband. In her Rajputana skirts and scarves of orange, salmon pink, buttercup yellow and emerald, below which appeared her small birdclaw ankles, twined with silver bangles, and her feet clad in a brand new pair of patent-leather dancing pumps, this girl, who did not appear to be more than sixteen years of age, who had never before even seen a motor-car, behaved with an innate elegance and calm, as she stood with pathetic simplicity, her draperies fluttering, and, lying in her upturned palm, the cross, cast from the cannon captured at Sebastopol. The hero's son, in a large blue and red turban, yawned and waggled a long stick while the recitation was made of her husband's heroic deeds of bravery. I could not refrain from making invidious mental comparisons with the gauche manner in which many English village girls might have deported themselves under such extraordinary circumstances.

The number of holidays in India surprised me. Although Hindus and Moslems hate each other, they enjoy each other's holidays. During the Holi, the Hindu spring, when for three days factories and shops are shut, the villages become a conflagration of colour and resemble more than ever a ballet. Many faces are covered with bright red paint — though one boy elects to wear a mauve face; eggs are filled with dyes and are hurled through the air; for months afterwards clothes bear witness to the festivities. During the festival of the Mohurran, edifices of tinsel and coloured paper are paraded through the streets, and everyone wears his most brilliant garment. Throughout the year the pictorial treats and surprises continue.

Sometimes the intense sun at midday creates an unbecoming lack of shadow, but as for the varying lights of India making of photography a "tricky business" — well, one can encounter every gradation of light by moving just a few paces in various directions, in and out of doors, on a summer's day in London. Processing, however, I found rather a problem, and not infrequently newly developed negatives were hung to dry by a window where a direct blast of sandladen wind would attack them. Only on my return to England and my faithful developers was I able to realize the true potentialities of some of my negatives. The enlargements of these pictures (all taken with a small camera) are, one year later, now coming in to remind me of the pleasant months of intensive travel that took me through most of the vast Peninsula, from Hyderabad, Deccan, to the North-West Frontier Province, from the mossy coolness of Simla to the sweltering villages of Bengal, from the coral City of Jaipur to the holy Mecca of Benares. How could my elderly Scot consider that photography in India could be anything but a joy, for no country affords greater opportunities?

C.B.

Village idyll

Bengali village scenes

Bengali labourer

Coolie boy, Bengal

Cattle grazing

BENGAL

Tilling a ricefield

Women trampling rice to make biscuits

Boy spinning in a Bengal village school

Thatched huts in a Bengal village

Writing lesson, village school

Punjabi village school class

Mission school children, Saraisha, Bengal

Rice cultivation, Bengal

Coolies carrying bundles of straw used for protective covering of ricefields

PESHAWAR

Fruiterer

Jeweller

Hand printer

Old rag merchant, Jama Masjid Mosque, Old Delhi

Girls' school, Bengal

Girls' school, Bengal

School children in class

Schoolgirl

Scenes in an ordnance factory, Calcutta

Jute factory, Bengal

Tochi scouts policing frontier

THE NORTH WEST FRONTIER

Sikh recruits: the swearing-in ceremony

The Khyber Pass

Gurkhas of the Fourteenth Army, Arakan

Jemaldar Bombardier Gurune, Arakan

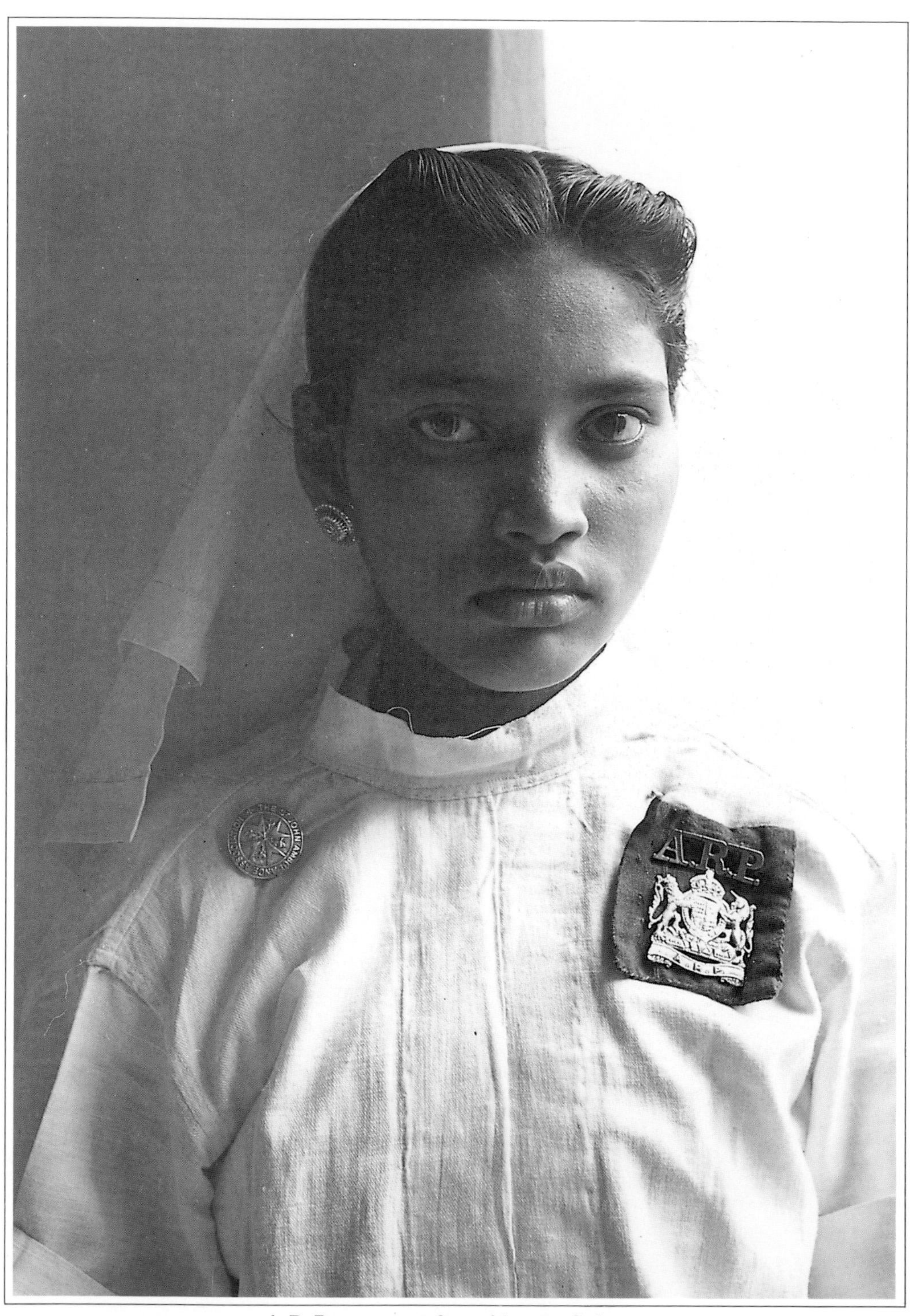

A.R.P. nurse at first-aid post, Calcutta

Carrying wounded from the Arakan jungle

Punjabi Mussulman Petty Officer, Royal Indian Navy

Indian naval officers in training

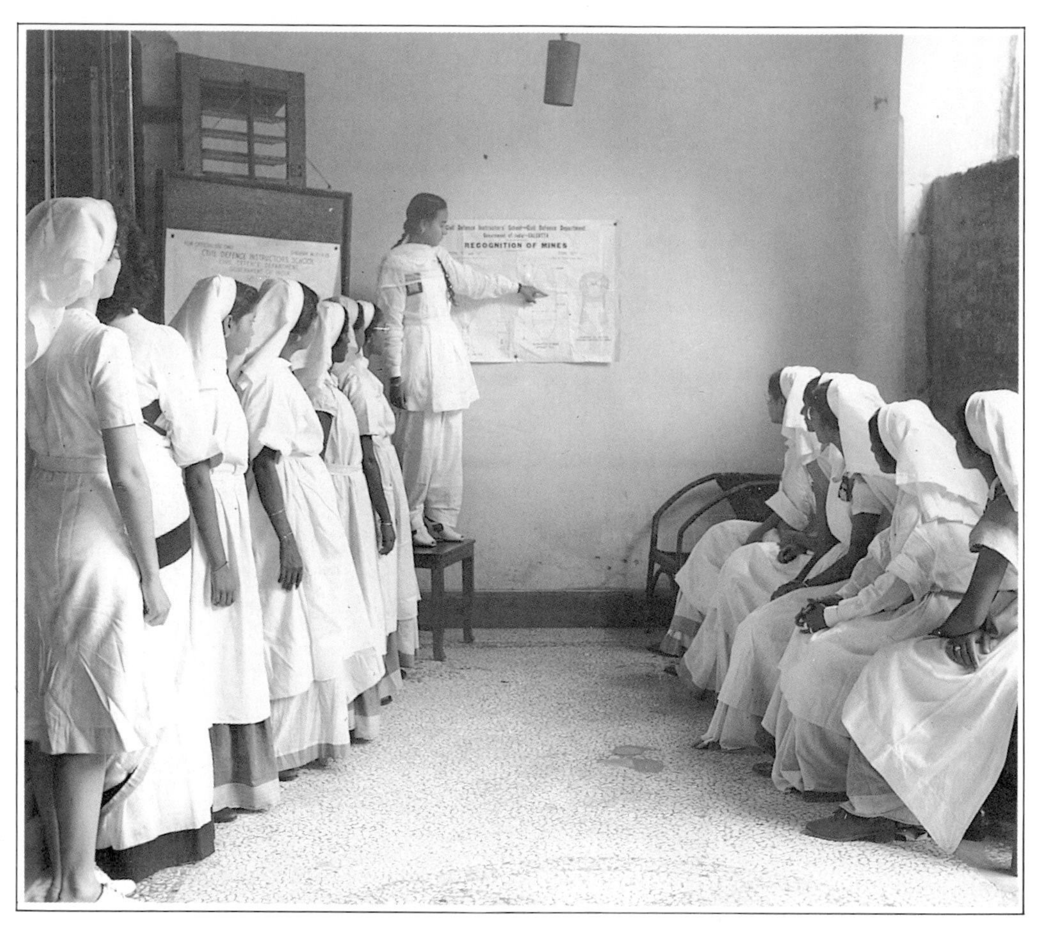

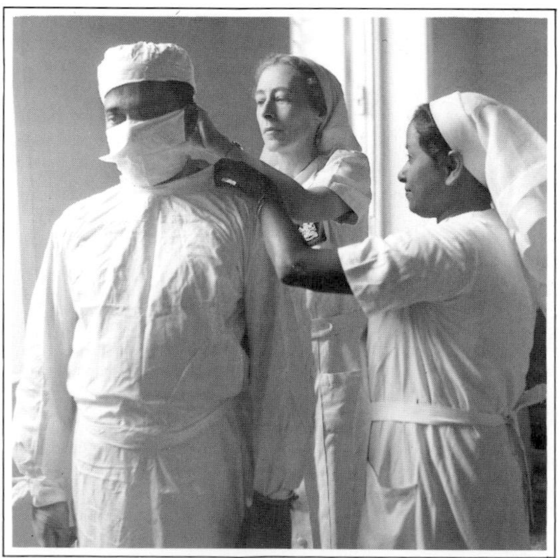

A.R.P. First Aid Post, Calcutta

A shoe-shine on Chowringhee, Calcutta

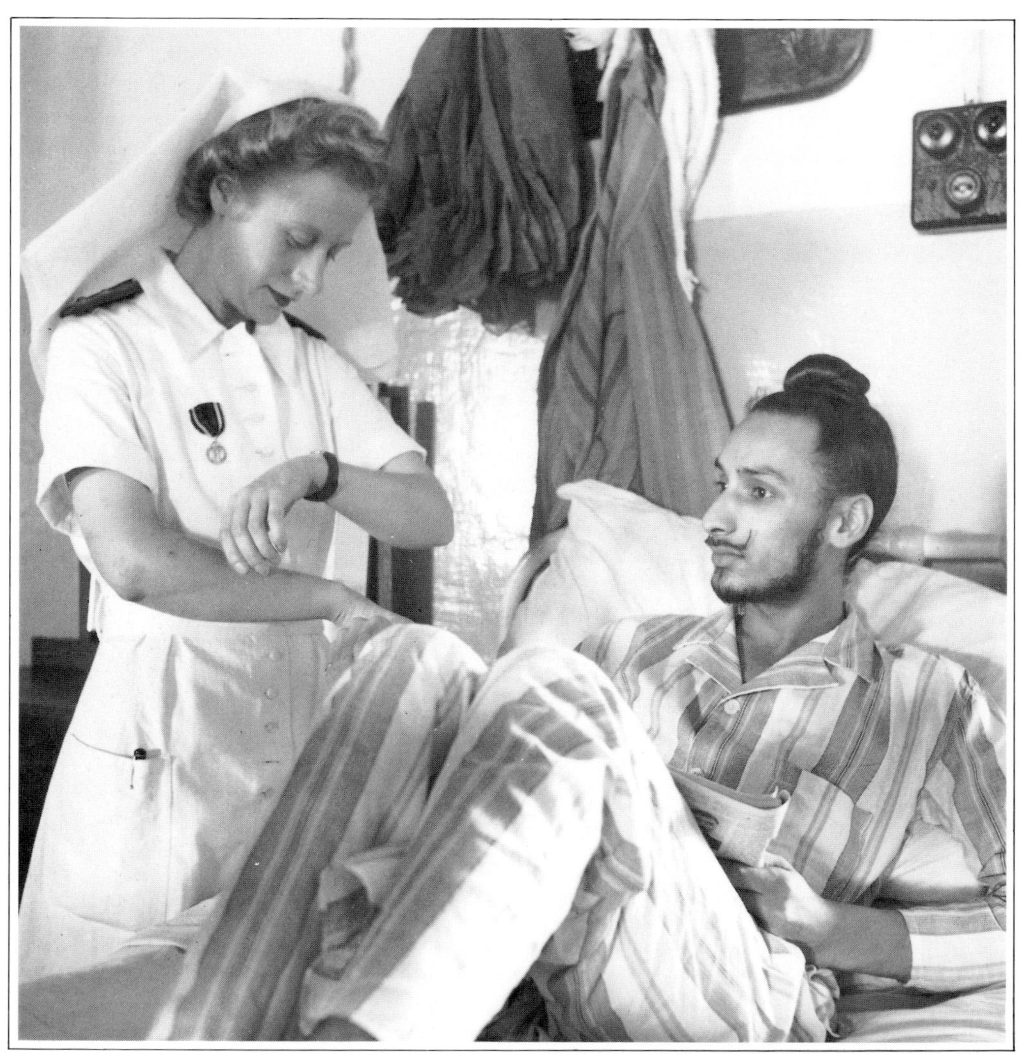

In the officers' ward, Military Hospital, Coiaba, Bombay

Bombay Ordnance Depot: landing craft in storage

Armoured vehicles in storage park

Beach Candy Bathing Club, Bombay

Mountain coolie

Coolies carrying wood for bridge-making along a mountain pass in the Arakan. The canvas screen is to prevent vehicles from going over the five hundred foot precipice

Stokers' mess deck, Royal Indian Naval Station, Sutlej, Calcutta

Punjabi

Ram Ghopal: Indian dancer

A pupil of Ram Ghopal dancing on the shore of the Arabian Ocean

Schoolgirls in class, Victoria Institution, Calcutta

Employee in a mathematical instrument factory in Calcutta calibrating a thermometer

Jaipur: Amber Palace

Mirror and stucco decoration, Amber Palace, Jaipur

Hyderabad City, Deccan

The sacred city of Benares

Students (and opposite) at Osmania University, Hyderabad

Dyers in a Jaipur Street drying saris in the sun

Boys in a temple courtyard

Mecca Masjid Mosque, Hyderabad

The Princess of Berar

The Maharanee of Jaipur

Carrion crows and pariah dog, Chowringhee, Calcutta

Temple in shopping centre, Benares

Bird market, Hyderabad

Kalighat: Hindus at the Temple of Kali

Hindu temple, Golconda Fort

Golconda, Deccan

Dyers' designs, Jaipur

The stupa, Sarnath, Benares

The Fort of Golconda

The sculptured caves of Ellora

Temple courtyard, Kailash cave, Ellora

Carved pillars and lions, Ellora

Sculpture in Kailash Temple, Ellora

Buddhist carving, Ellora

Eighteenth-century tombs of English pioneers, Calcutta

The garden of the Red Fort, Old Delhi

Inside the Red Fort, Old Delhi

Jama Masjid Mosque, Old Delhi

Wall paintings in Ajanta cave

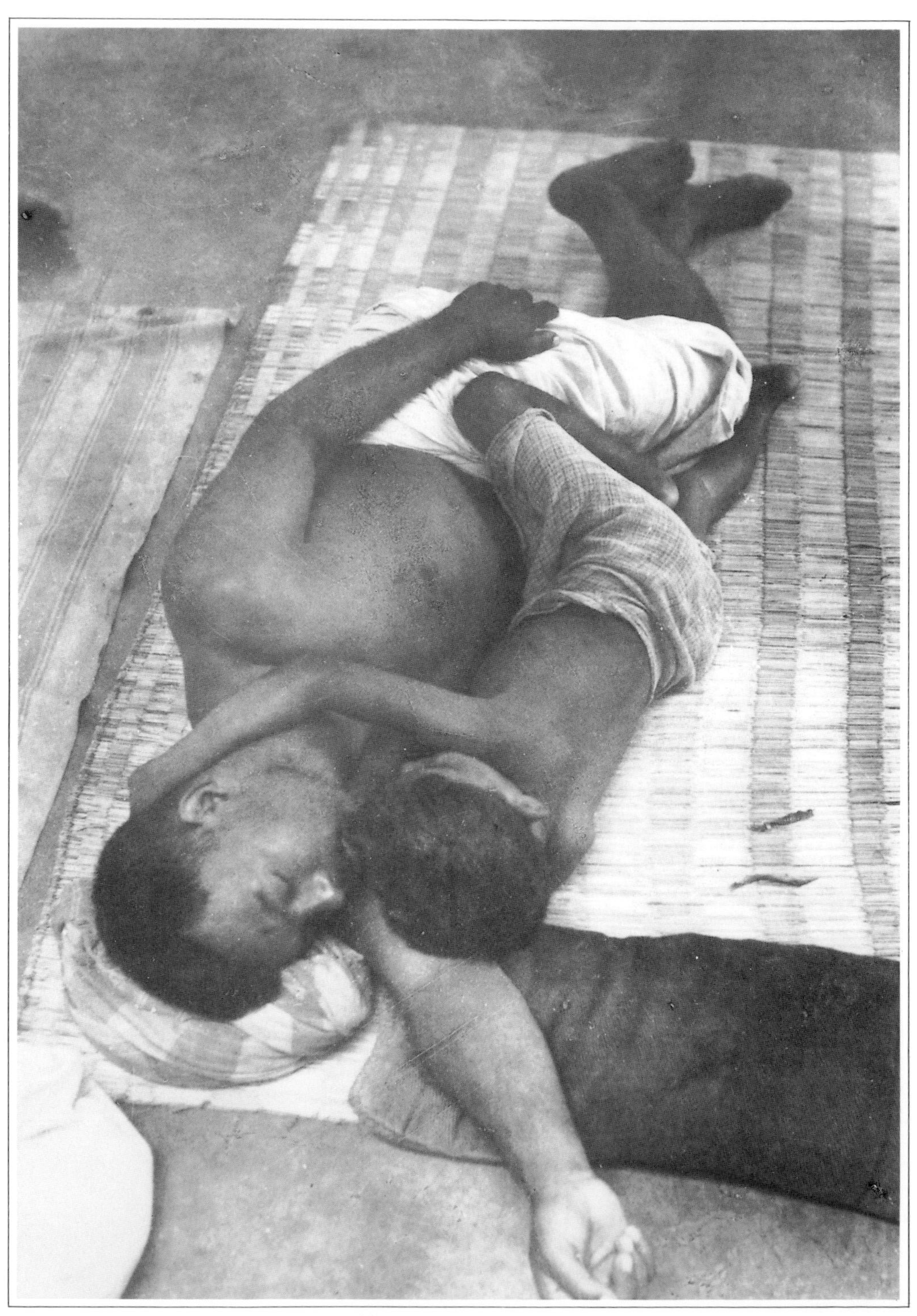

Siesta: father and son

Waiting: mother and daughter

The garden, Jain Temple, Calcutta

Eighteenth-century mosque, Hyderabad City

Glass inlay decoration, Calcutta

Marble inlay decoration, Red Fort, Old Delhi

Roadside scene, Bengal

Fish-net weaver

Village child, Bengal

Village school child

Bengali village boy